Regional Political Ecologies and Environmental Conflicts in India

This book focuses on the regional political ecologies (RPEs) of environmental conflicts in India. It broadly explores landscape-based analyses of political, economic and social issues which impact environmental changes, challenges and conflicts at local and micro-local levels.

The chapters in this volume examine the intervention of different stakeholders in the management of various regional ecological landscapes in India, including forests, rivers, canals, creeks and wetlands. The volume is an interdisciplinary endeavour, weaving together contextual narratives through a combination of approaches from sociology, anthropology, geography, political studies and environmental history. Using such core approaches, the book studies the place-based dynamisms within the regional environmental conflicts in the selected conservation landscapes. It provides empirical reflections on transboundary issues, rural–urban transitions, middle-class environmentalism, identity conflicts, decentralized natural resource management and the role of political institutions.

Regional Political Ecologies and Environmental Conflicts in India will be of great interest to students and scholars of Political Ecology and South Asian Environmental Studies.

Sarmistha Pattanaik is Associate Professor at the Department of Humanities and Social Sciences and Associate Faculty of Centre for Policy Studies (CPS) and Centre for Urban Science and Engineering (C-USE), Indian Institute of Technology Bombay, Mumbai, India.

Amrita Sen is Assistant Professor at the Department of Humanities and Social Sciences, Indian Institute of Technology, Kharagpur, West Bengal, India.

Routledge Focus on Environment and Sustainability

Water Governance in Bolivia
Cochabamba since the Water War
Nasya Sara Razavi

Indigenous Identity, Human Rights, and the Environment in Myanmar
Local Engagement with Global Rights Discourses
Jonathan Liljeblad

Participatory Design and Social Transformation
Images and Narratives of Crisis and Change
John A. Bruce

Collaborating for Climate Equity
Researcher–Practitioner Partnerships in the Americas
Edited by Vivek Shandas and Dana Hellman

Food Deserts and Food Insecurity in the UK
Exploring Social Inequality
Dianna Smith and Claire Thompson

Ecohydrology-Based Landscape Restoration
Theory and Practice
Mulugeta Dadi Belete

Regional Political Ecologies and Environmental Conflicts in India
Edited by Sarmistha Pattanaik and Amrita Sen

For more information about this series, please visit: www.routledge.com /Routledge-Focus-on-Environment-and-Sustainability/book-series/RFES

Regional Political Ecologies and Environmental Conflicts in India

**Edited by Sarmistha Pattanaik
and Amrita Sen**

LONDON AND NEW YORK

First published 2023
by Routledge
4 Park Square, Milton Park, Abingdon, Oxon OX14 4RN

and by Routledge
605 Third Avenue, New York, NY 10158

Routledge is an imprint of the Taylor & Francis Group, an informa business

British Library Cataloguing-in-Publication Data
A catalogue record for this book is available from the British Library

ISBN: 978-0-367-48642-6 (hbk)
ISBN: 978-1-032-41782-0 (pbk)
ISBN: 978-0-367-48643-3 (ebk)

DOI: 10.4324/9780367486433

Typeset in Times New Roman
by Deanta Global Publishing Services, Chennai, India

Contents

Contributors

Radhika Bhargava, PhD scholar, Department of Geography, National University of Singapore.

Shreyashi Bhattacharya, PhD scholar, Rekhi Centre of Excellence for the Science of Happiness, Indian Institute of Technology Kharagpur.

Lina Bose, PhD scholar, Department of Humanities and Social Sciences, Indian Institute of Technology Kharagpur.

Sana Huque, Post-Doctoral fellow at ATREE, Bangalore and a former Research Associate with Environment Support Group, Bangalore, India.

Janmejaya Mishra, PhD scholar at Ashank Desai Centre for Policy Studies (ADCPS), Indian Institute of Technology Bombay, India.

Jenia Mukherjee, Assistant Professor, Department of Humanities and Social Sciences, Indian Institute of Technology Kharagpur, India.

Avijit Sahay, Post-doctoral scholar, Department of Humanities and Social Sciences, Indian Institute of Technology, Bombay, India,

A. Wati Walling, Assistant Professor and Head, Department of Science and Humanities (Sociology), National Institute of Technology (NIT) Nagaland, Chumukedima, Dimapur, Nagaland, India.

Acknowledgments

The editors would like to sincerely thank all the contributors for their insightful research, open engagement and timely communication of their chapters to this edited volume. At the same time, we appreciate the patience of all the contributors for their support – starting from the conception to the final submission. This edited volume would not have been a finished one without the expertise of these stellar researchers.

Many thanks to the entire Routledge team – it has been a joy indeed to work with them. We would like to sincerely thank the team and their associates for their professionalism, continuous support, patience and excellent coordination throughout different stages of development of this volume: particularly Matt Shobbrook, Editorial Assistant.

Finally, we are grateful for all the comments received from the potential reviewers at the time of review, which helped us to make substantial improvements to this volume. However, we take full responsibility for the deficiencies.

Sarmistha Pattanaik
Amrita Sen

1 Introduction

Understanding Regional Political Framings of Environmental Conflicts in India

Sarmistha Pattanaik and Amrita Sen

Analyzing environmental conflicts in India necessitates attention to multiple ways in which claims to natural resources are established by a range of sub-national and global actors. Many such conflicts, being regionally positioned, render an appropriate scope of applying a framework of regional political ecology (RPE) to incorporate an understanding on scalar political contexts of human-inhabited landscapes. RPEs can be significant in explaining landscape-based analyses of the interplay of political, economic and social factors influencing environmental conflicts in the current decades, characterized by newer institutional approaches to environmental governance (Paavola, 2007; Wilshusen, 2019; Kashwan, McLean & García-López, 2019). Introduced by Blaikie and Brookfield in 1987 in their book *Land Degradation and Society*, RPE implied a 'broad based approach encompassing a variety of scales, methodologies and conclusions regarding the cause of land degradation' (Black, 1990: 35–36). It was an epistemic attempt to look into environmental crisis (land degradation in the specific case) as manifested by plural and regionally specific processes, accounted in the wider political economy of natural resource management. In this volume, the intention is to highlight how regional approaches and the utility of specificities 'scale-up' rational generalizations on the political construction of ecologies across the globe.

Socio-environmental and political changes are contextually specific, and despite having a geographical coherence, they are often in a state of ambiguity and contestability, urging an explicit independent inquiry (Walker, 2016: 125). Region, as Galt (2016) points out, matters in the making of this inquiry, making its way through binaries in exploring bounded territories with distinct political ecological imageries. RPE aims to transcend representative global framings, which are often applied on local fields poorly positioned to contextualize such broad-based comparative analysis (Walker, 2003: 7). It also aims to apply inter-regional political framings on meso-geographic

DOI: 10.4324/9780367486433-1

scales, enabling a representation of 'distinctiveness of regions' coexisting with 'commonalities of individual places' (Walker, 2003: 21–22). Miller and McGregor (2019: 663) point towards the usefulness of 'world regional approaches' to identify analytical and political approaches in explaining supra-national processes – 'comparative analysis of place-based, single issue research; generation of diverse counter-narratives at the regional scale; and consideration of flows and networks'. A place-based political ecology approach, they argue, complements traditional concerns on equity, justice and marginalization in enabling an exploration of the politically dominant forces at scales in shaping environmental agendas (Miller & McGregor, 2019: 673). In more recent essays on the centrality of 'regions' in political ecology and environmental justice analysis, scholars identify the need to rethink place-based inequities and actions to address them (London, 2016).

Political ecology studies the power dynamics and inequities in natural resource management access across groups and individuals (Forsyth, 2003; Robbins, 2012). From its emergence in the 1970s, political ecology explained the interplay between ecology and political economy in diverse contexts (Gadgil & Guha, 1995). It underscores the fact that inequality, power asymmetries and political actions influence environmental governance at multiple levels. In this volume, we argue that a broader and region-based approach in understanding political ecologies prompts a deeper sociological analysis of discrete conservation landscapes in emerging contexts (Moore, 1998; Murphy, Enqvist & Tengo, 2019; Sen & Nagendra, 2019, 2020). Pioneering discourses on environment agendas are mostly sketched through binary imageries positioning 'first' and 'third' worlds as representative geographic frames in political ecology (Walker, 2003: 8) – a glance on the trajectories of Northern and Southern environmentalism would better contextualize such visions on environmental agendas. While Northern ideologies arose from a consciousness garnered over widespread environmental degradation, in the global South, threats to ecology and denial of rights to livelihood, fuelled by commercial exploitation of natural resources led to environmental movements. Guha and Martinez-Alier (1997) illustrated a distinction between 'full stomach environmentalism' of the global North and 'empty belly environmentalism' of the global South – they explained how environmental movements in the South are fostered by threatened livelihoods, owing to environmental degradation, contrary to the North, where 'people-free' ecological sites are preferred as ideal to conservation. Indigenous knowledge and its merits in conserving ecosystems have been demonstrated sufficiently strongly in Southern environmentalism (Gadgil, Berkes & Folke 1993). In Southern environmentalism, 'issues of ecology are interlinked with questions of human rights, ethnicity and distributive justice' (Guha & Martinez-Alier, 1997: 18).

However, in both Northern and Southern forms of environmentalism, a watertight compartmentalization of ecological habitats, as shaped by homogenous social-ecological association, precludes an understanding of nature characterized by a 'porosity of boundary' (Escobar, 2001: 144). Such framings of environmentalism typically represent attachments to natural spaces as either centred around conservation of 'wild' and 'pristine' ecological landscapes or as harnessed by indigenous rural communities with longstanding, older and sedentarized associations to place (Sen & Nagendra, 2020: 1). Recent studies underscore the significance of environmental placemaking in conserving nature – collaborative and synthesized approaches, which can contribute synergistically to nature conservation (Seeland, Dübendorfer & Hansmann, 2009; Byrne, Goodall & Cadzow, 2010; Sen & Nagendra, 2019, 2020). These studies explain that in a fast-changing world, indigenous and regionally specific histories of socio-cultural attachments to landscapes are not the only factors leading to conservation initiatives. In other words, such attributes are not binding or essential for establishing place identities with nature. Environmental placemaking prompts readers towards recognizing ecological conflicts as chequered and non-linear, shifting discourses towards capturing complexities within place-based framings on the impact of conservation norms on diverse stakeholders. Collaborative approaches can be instrumental in addressing regional environmental challenges that natural habitats undergo.

A RPE model allows an explicit focus on the operation of regional networks, prompting environmental management and animating social-ecological processes within particular regions, thus upscaling the analytical applicability of these processes to other regions as well – 'ensuring that it is these processes, rather than the scale itself that remain in analytical focus' (Miller & McGregor, 2019: 676). The reason why scales make a difference is because they point attention towards meso-political and economic forces, shaping relationships between nature and society in ecologies with sharply disaggregated social interests. This is imperative, considering large-scale transformations in the nature of environmental policies in the recent decades and a growing public participation in processes designed towards environmental protection. However, increasing participatory powers and empowerment policies, mandated within recent management frameworks, fail to reduce socio-economic marginalization to a large extent. With current policies largely transforming landscapes through globally pervasive environmental governmentality and globally produced and valued eco-knowledge, human relationships with nature are likewise transitional and evolving (Goldman, 2001; Bayon & Jenkins, 2010). A range of scholarship on Southern environmentalism, engaging in a political ecology framing, explains how local claims to resources are often translated into political choices through mechanisms of 'political intermediation' – one that is shaped through different means like politically engaged movements, politically structured advocacies and are effected through successful mechanisms of representation at the national, sub-national and local levels (Kashwan, 2017: 13–16). RPEs explain how integral are the processes of 'legitimizing' rights in prioritizing human spaces amidst globally recognized wilderness landscapes. This is because mobilization towards rights is not always politically neutral – a political ecology analysis helps understand that it is powerful institutions and political actors which represent distinctive agendas on the legitimation rights as collective priorities (Menon & Karthik, 2019; Sen, 2022).

Based on a collection of case studies, this volume introduces six essays on the RPEs of environmental conflicts in India. To represent environmental conflicts in a more nuanced way, this topical volume focuses on different forms of ecologies such as forests, rivers, canals, creeks, wetlands and uses an RPE approach to examine key issues of politics, power, justice and gender in constructing a 'chain of explanations' defining broader challenges (Blaikie & Brookfield, 1987: 239). Such a chain of explanations offers regional cases explaining wide arrays of politico-economic factors accounting for environmental challenges, contributing holistically to conceptual framings on the relationships between politics and the environment. Despite foregrounding empirical evidence, the volume significantly renders scope for advancing the epistemological debate on the field of RPE. This is primarily because the volume is an interdisciplinary endeavour, weaving together contextual narratives through a combination of approaches like sociology, anthropology, geography, political studies, policies and environmental history. Using such core approaches, the volume studies dynamisms within regional environmental conflicts in the selected conservation landscapes of India. It provides empirical, place-based reflections on transboundary issues, rural–urban transitions, middle-class environmentalism, identity conflicts, decentralized natural resource management and the role of political institutions. As a part of this exploration, the volume will holistically explore ways in which regional environmental conflicts can be largely 'politicized' and manifest in material transformations of socio-ecological relations. While exploring the politicization of environmental conflicts, our attempt has also been to show how multiple and cross-cutting agendas of justice and environmental concerns are articulated across politically animated conservation landscapes. Individual contributions will provide innovative and original epistemological advances on RPE and scope for indicating new agendas in academic research, policy as well as activism.

The uniqueness of this volume lies in its intersectionality of perspectives in understanding the RPE framing of environmental conflicts.

It complements the ongoing scholarship in multiple ways: (1) it employs a cross-disciplinary approach to understand human–nature relationships and how ideas of justice, rights and identity shape the regional politics of environmental conflicts; (2) it aims to understand the current environmental regulations and the place-based conflicts around the interpretations and implementation of such regulations; (3) it presents some of the most contemporary and complex challenges in addressing the regional environmental conflicts, given the fact that all the chapters present complex social issues manifested within those conflicts; (4) it represents cross-disciplinary perspectives on the environmental conflicts of regional landscapes and provides advances in the global political ecology discourse. The aim of this volume is to provide an alternative version of political ecology – according to Walker (2003: 8),

> regional approaches can retain the greatest strengths in revealing the importance of local-scale social dynamics, while situating these dynamics within broader scales of regional (and global) processes – providing greater coherence while avoiding such problematic frames as the 'first' and 'third' world.

The volume would help in framing contextual political ecology perspectives, and situate them within the broader epistemological debate on global political ecology.

The diverse analytical frameworks offered in the individual chapters would help enhance understanding on RPEs of environmental conflicts, along with illustrating contemporary forms of post-structuralist and neoliberal environmental conflicts at the meso level. After this Introduction, in Chapter 2, the author explores reflections on the politics of environmentalism surrounding the conservation of Adi Ganga, a canal of the Ganges that runs through the city of Kolkata, demonstrating different shades of environmental conflicts and the politics of protests across multiple actors – this is analytically explained using a political and plural perspective on competing environmental agendas. The chapter discerns politics of pollution, petitions and 'varieties' of protests, highlighting diverse cultural preferences on ecological conservation in a city. While radical grassroots protests remained restricted to the direct correlation between the state of the River and basic survival provisions for the urban subaltern, a 'heritage' argument positioned in a middle-class environmentalism helps explain the plight of Kolkata's educated, cultured and affluent activists on urban nature – people to whom the rich riverine history of the Adi Ganga and riverine flows connecting Kolkata city within its larger deltaic-estuarine scape seemed to be extremely significant. The final outcome of the chapter is not only an exposition on power hierarchies but also an attempt to offer reflections on how complex political histories of cities would shape future urban sustainability.

Chapter 3 looks at the importance of borders in generating multiple social, cultural, economic as well as ecological opportunities on a global scale. This chapter also discusses possible challenges to attaining these opportunities while looking for solutions to effective management from the field of adaptive governance. It is highlighted that a transboundary conservation approach is especially interesting due to its multidimensional social, political and ecological benefits, thus, the challenges that are required to overcome for the realization of these opportunities are poor communication, weak governance and high costs. This chapter focuses on understanding the border in the context of transboundary conservation of natural resources. The two questions this chapter answers are: (1) how does a border necessitate transboundary conservation? and (2) under what circumstances can transboundary conservation be effective?

In Chapter 4, reflections on the political implications of riverbank erosion in the Majuli Island of Assam offer conceptual and empirical insights on how place-specific environmental crises manifest in transformations of social associations in ecologically threatened landscapes. The author explains how riverbank erosion by River Brahmaputra in the Majuli Island of Assam has led to the displacement of thousands of people and rendered them landless. The erosion has also altered relationships between the dominant Assamese community and the minority Mising tribe of the island. The Mising are a fishing community that inhabits the riparian zone of River Brahmaputra. The culture of Majuli Island is dominated by neo-*Vaishnavism*, a social and cultural movement of the Assamese community founded by Shankardeb in the 15th century. While the Mising practice many elements of neo-*Vaishnav* culture, they have never been fully accepted or assimilated into the neo-*Vaishnav* fold by the dominant Assamese community. This unequal distribution of power and space is further reinforced by riverbank erosion, necessitating a RPE analysis. Being a fishing community, the Mising are disproportionately affected by the changing course and erosion of Brahmaputra. Loss of land and fishing grounds, coupled with the unequal distribution of political and geographical space in the rest of the island means that many in the community are resenting and challenging the neo-*Vaishnavite* hegemony of the island. The present paper analyzes a nuanced and intricate politics of sharing the cost of riverbank erosion among the different communities of Majuli and the ecological and cultural implications of it.

In Chapter 5, an analysis of the conservation challenges in East Kolkata Wetlands is presented by a RPE of landscape transformations, explaining effects on the health of urban ecology as well as the provisioning services of the marshland. The wetlands are often referred to as the 'Kidneys of Kolkata' because of the unique mechanism they possess in naturally treating and purifying waste water. However, in recent times, the area has come under repeated onslaught of conversions, both legal and illegal, and encroachments. In addition, the current government and a large section of the urban inhabitants, as well as the wetland communities who are increasingly veering towards consumerism, exclusivism and divisiveness, see it as an empty space into which the city can expand its urban footprints. Using a RPE perspective, this paper is an attempt to examine impacts of the new urban aspirations of Kolkata in a protected site like the East Kolkata Wetlands. The concept of political ecology increasingly strives to integrate political economy with the growing concerns about the health of ecology. So at a time when the city is witnessing a form of urbanization that has become characteristic of neoliberalism-driven economic growth, RPE can offer an explanation for the reason and manner behind the reclamation and straightjacketing of the East Kolkata Wetlands into land for the urban market. Using political ecology, the paper shall attempt to analyze the role of the state and its agencies in the transformation of the wetlands and analyze the nature of response from the people of the city and the wetlands towards the dismantling of a protected site like the East Kolkata Wetlands.

Chapter 6 reveals empirical observations from land ownership patterns in Nagaland and significant merits for a nuanced RPE analysis of natural resource management. Nagaland has an unusual land ownership pattern with about 88% of land being owned by either individuals or tribal communities and a meagre 12% owned by the state. Rightly so, much has been said about the region's customary land ownership practices and its colligated natural resource (mis)management, underdevelopment, conflict, resource scarcity, poverty, etc. However, beneath such claims there are multi-layered and intersectional components, such as ecological knowledge production, power and gender perspectives, which are rather complex and beyond the obvious reasons attributed. This chapter analyzes a much-ignored perspective of a politicized universe (a Naga village here), where gender and power play consequential and significant roles in land ownership and natural resource management. The paper would argue that North East India and its extended territories like Nagaland need a nuanced perspective such as political ecology to examine the geopolitics of the region as well as the complex interplay between gender and power in the ownership of land and resources.

Finally, Chapter 7 critically examines the contemporary processes of governance of mineral and forest resources in the Bastar division of Chhattisgarh. Drawing from the long-term ethnographic research by the author, this chapter reflects on the regional political ecology of natural resource appropriation deeply entrenched in the neoliberal agenda of the state which has significantly deprived the Scheduled Tribes of their fundamental rights and made their livelihoods highly vulnerable. The over-extraction of natural resources such as minerals and forests, the exclusionary development processes – establishment of national parks and proposed hydro-power projects – has largely benefitted the Indian state and a few private corporations at the cost of marginalization of the Scheduled Tribes in the region. The chapter suggests that effective implementation of public policies such as the Panchayats Extension to Scheduled Areas Act, 1996 and the Forest Rights Act, 2006, efficient utilization of the royalties from mining activities in the socio-economic development, strengthening the forest-based livelihoods and ensuring active participation in the governance of natural resources will safeguard the rights, conserve the rapidly degrading resource base, provide sustainable livelihood opportunities and enhance the well-being of the Scheduled Tribes by putting an end to the ongoing resource conflict in Bastar, Chhattisgarh.

The fundamental theme that binds all the chapters together is their focus on the interplay between regional conflicts and environmental politics. All the chapters distinctively focus on this interplay, through a RPE analysis that is place-based and merits positionality within the larger framework of understanding. It presents regional environmental issues of relatively unexplored locations in the geographical landscape of India, yet doesn't lose sight of its ability to contextualize its findings within a coherent analytical framework of political ecology. This is fundamentally so, since the chapters do not merely present a binary while speaking about environmental conflicts, but simultaneously discuss discrete issues of power, inequality and recognition in their individual analyses. The volume would also be an important addition to the academic discourse underscoring environmentalism of the global South. It offers a pan-regional approach in understanding nuances that are constitutive of the definition of RPE.

References

Bayon, R., & Jenkins, M. (2010). The business of biodiversity. *Nature, 466*(7303), 184–185.

Black, R. (1990). Regional political ecology in theory and practice: A case study from Northern Portugal. *Transactions of the Institute of British Geographers, 15*(1), 35–47.

Blaikie, P., & Brookfield, H. (1987). *Land degradation and society*. London and New York: Routledge.

Byrne, D. R., Goodall, H., & Cadzow, A. (2010). *Place-making in national parks: Ways that Australians of Arabic and Vietnamese background perceive and* use the parklands along the Georges River. NSW. Retrieved from http://www.environment.nsw.gov.au/ resources/cultureheritage/OEH20120073PlaceMaking.pdf.

Escobar, A. (2001). Culture sits in place: Reflections on globalism and subaltern strategies of localization. *Political Geography, 20*(2), 139–174.

Forsyth, T. (2003).*Critical political ecology: The politics of environmental science.* London: Routledge.

Gadgil, M., Berkes, F., & Folke, C. (1993). Indigenous knowledge for biodiversity conservation. *Ambio, 22,* 151–156.

Gadgil, M., & Guha, R. (1995). *Ecology and equity: The use and abuse of nature in contemporary India.* London and New York: Routledge.

Galt, R. E. (2016). The relevance of regional political ecology for agriculture and food systems. *Journal of Political Ecology, 23*(1), 126–133.

Goldman, M. (2001). Constructing an environmental state: Eco governmentality and other transnational practices of a 'green' World Bank. *Social Problems, 48*(4), 499–523.

Guha, R., & Alier, J. M. (1997). *Varieties of environmentalism: Essays North and South.* London: Earthscan.

Kashwan, P. (2017). *Democracy in the woods: Environmental conservation and social justice in India, Tanzania and Mexico (studies in comparative energy and environmental politics).* New York: Oxford University Press.

Kashwan, P., Maclean, L. M., & García-López, G. A. (2019). Rethinking power and institutions in the shadow of neoliberalism (an introduction to a special issue of World Development). *World Development, 120,* 133–146.

London, J. K. (2016). Environmental justice and regional political ecology converge in the other California. *Journal of Political Ecology, 23*(1), 147–158.

Menon, A., & Karthik, M. (2019). Genealogies and politics of belonging: People, nature and conservation in the Nilgiri Hills of Tamil Nadu. *Conservation and Society, 17*(2), 195–203.

Miller, F. P., & McGregor, A. (2019). Rescaling political ecology? World regional approaches to climate change in the Asia Pacific. *Progress in Human Geography, 44*(4), 663–682.

Moore, D. S. (1998). Subaltern struggles and the politics of place: Remapping resistance in Zimbabwe's Eastern Highlands. *Cultural Anthropology, 13*(3), 344–381.

Murphy, A., Enqvist, J. P., & Tengö, M. (2019). Place-making to transform urban social ecological systems: Insights from the stewardship of urban lakes in Bangalore, India. *Sustainability Science, 14*(3), 607–623.

Paavola, J. (2007). Institutions and environmental governance: A reconceptualization. *Ecological Economics, 63*(1), 93–103.

Robbins, P. (2012). *Political ecology: A critical introduction.* Oxford: John Wiley and Sons.

Seeland, K., Dubendorfer, S., & Hansmann, R. (2009). Making friends in Zurich's urban forests and parks: The role of public green space for social inclusion of youths from different cultures. *Forest Policy and Economics, 11*(1), 10–17.

Sen, A., & Nagendra, H. (2019). The role of environmental placemaking in shaping contemporary environmentalism and understanding land change. *Journal of Land Use Science*, *14*(4–6), 410–424.

Sen, A., & Nagendra, H. (2020). Local community engagement, environmental placemaking and stewardship by migrants: A case study of lake conservation in Bengaluru, India. *Landscape and Urban Planning*, *204*, 103933.

Sen, A. (2022). *A political ecology of forest conservation in India: Communities, wildlife and the state*. Abingdon and New York: Routledge.

Walker, P. A. (2003). Reconsidering regional political ecologies: Towards a political ecology of the rural American West. *Progress in Human Geography*, *27*(1), 7–24.

Walker, P. A. (2016). On 'Reconsidering Regional Political Ecologies' 13 years on. *Journal of Political Ecology*, *23*(1), 123–125.

Wilshusen, P. R. (2019). Environmental governance in motion: Practices of assemblage and the political performativity of economistic conservation. *World Development*, *124*, 104626.

2 Heritage or Basic Human Rights?

Politics of Environmentalism Surrounding the Adi Ganga in Kolkata

Jenia Mukherjee, Shreyashi Bhattacharya and Lina Bose

2.1 Introduction

2.1.1 Rivers, Rights and Heritage

Adi Ganga, the river that witnessed the transformation of the Lower Gangetic Delta from a swampland to a megapolis, succumbed to various anthropogenic activities that completely disregarded the ecological sensibilities of a natural resource in the post-colonial period. Once a distributary of historical significance, it now lies submerged in filth and is defunct in multiple spaces, sparking numerous conversations from grassroots to policymakers on the need for river rejuvenation. But what makes a structure, whether natural or man-made, deserving of protection and nurturement beyond economical distinctions? Adi Ganga, or any other river, is much more than only her financial attributes. Recent discourses on the river's revitalization and conservation process focus mainly on renewing its beneficial services for the city of Kolkata, taking the conversation away from the increasing nature-culture dichotomy across multiple stakeholderships. A river cannot be the same for all but all are the same for the river. So many narratives provide a fluid perception of the river which is as dynamic as its physicality but also held together by centuries of heritage.

In the heritage discourse, the term usually classifies two types of heritage: tangible, ascribing to hard or solid infrastructures that can be touched or seen; and intangible heritage, referring to soft or evolving aspects of heritage that are passed on over time by associating it with knowledge. Although mainstream conservationists have always propagated the preservation of tangible forms of heritage, in recent times, debates on protecting intangible forms have also emerged. Harrison's (2013) observation on

DOI: 10.4324/9780367486433-2

critical heritage studies defines heritage as per its functionality. He defines heritage as ubiquitous, with the concept and meaning of the term to have expanded in such a way that everything (historic, architectural artefacts, old structures) is considered to be a heritage, putting forward the dialogical model that dissolves all hierarchies and integrates varied social actors, theoreticians who now simply see the past as distant from the present. However, the World Heritage Convention in the early 1970s suggested that heritage is not only about past but instead about our relationship with the present and the future. Similarly, the heritage of a river, which is often a transboundary resource, cannot be limited to its binary tangibility, but can be explored across the creative engagement between the past and present of human–nature relationship.

The story of Adi Ganga is another testimonial to the fact that its heritage lies not only in what it lost in time but the memories it has left the generations with and the civilization that it helped evolve on its banks. The differences in the residences sprouting across the banks of Hooghly at Dakhineswar and Adi Ganga at Kalighat, are another reference to the geocultural contribution of a river in the development of heritage in a community. However, the meaning and theoretical concept of heritage has shifted since then and is now seen as a process strictly linked with bureaucratic high modernist planning dealing with an integrated First World agenda of sustainable development while ignoring the varied Third World sensibilities of ecological access and justice among the margins. While the role of political ecology in understanding this interplay between ecological access and political economy and examining intersectionalities across power dynamics and inequalities in the context of environmental justice is unparalleled in the social science discourse, the case of Adi Ganga offers a much more grounded and focused opportunity to analyze the power equations along specific spatial and temporal scales through Regional Political Ecology (RPE) that can be helpful in identifying and exposing context specific plural realities, especially in Third World post-colonial societies.

The concept of RPE first emerged in the late 1980s as an approach to describe land degradation. Blaikie and Brookfield (1987) wrote about the integration of the human and the physical environment in their historical, political and economic context to understand the cause for the degradation. This application of a locality-based approach suggested the exploration of different power hierarchies in different societies, thus spinning the attention on the 'management' of the resources in different communities rather than separating the natural and man-made causes (Black, 1990). Blaikie's earlier works (1985) that focused on political economy and environmental changes identified the complex processes of environmental degradation where state intervention, or in some cases the management system, would lead to the degradation of the resources, causing the marginalized communities who depended on it to turn to the world economic system. So, the question that needs to be bluntly addressed is, why is 'region' important in political ecology, and where does it fit into decomplexifying the interactions between the local infrastructures and state-legitimized environmentalism?

In the field of political ecology multiple interpretations of the term have been addressed which suggest that political ecology explores ecological–political interdependencies, but Walker (2003) highlights the lack of understanding of nature–society's dynamic political economy and ecological interdependence on basis of scale across First and Third World nations. The blatant homogeneity of the framework across different ecologies in the world diffuses the integrated broader as well as micropolitics in the region. The effects of the anthropocene have been varied across nations and communities as a result of differing geographical as well historical and political attributes, therefore leading inadvertently to different social–ecological equations in contrasting regions. Robbins' (2002) argument of PE, however explored the challenges of the discourse which mostly focused on the equations between the local informal institutions rather the oppressive state machinery. While McCarthy (2002) showcased certain similarities among the First and Third World political ecology cases, without further elaboration it must be taken into account that, while the approach and issues of political ecology might be considered with Third World nations, the solutions that are offered are often straitjacketed with First World equations. Emphasizing on 'region' in political ecology might inverse the situation and provide some coherence on this subject by eliminating this First World/ Third world binary challenge.

It is through the scaling of the theoretical concept that the differences in global and Adi Ganga conservation can be traced. This can be understood through the simple example of the Whanganui River in New Zealand. Whanganui was the first river in the world to have received a legal human status as a result of the constant perusal of the Maori tribesmen, the true custodians of the river. The emphatic connection between the Maori and Whanganui River is reflected through their perception of the river, i.e., 'I am the river, the river is me'. However, in 2017 a similar verdict was passed by the Uttarakhand court in the context of River Ganga and Yamuna that lacked the emphatic connotations of the New Zealand verdict as the former's issue focused on access and that of the marginalized people. A similar but also different narrative is noticed in the case of Adi Ganga, where access and exclusion dominate the trajectory of the environmentalism and political economy of the resource. The following empirical section allows one to understand why, despite being called the Narmada of Kolkata, activism surrounding Adi Ganga or the Tolly's Nullah didn't follow a similar response as of the former's campaign and success, or of the Whanganui, but instead revealed the multifaceted dynamism of heritage in the context of ecological urbanism through the lens of RPE.

2.2 Case Study: Adi Ganga

2.2.1 Recalling a Forgotten River

Mukherjee's (2020) account of Kolkata's 'blue infrastructures' captures the 'modern' departure from 'ecology of cities', an erroneously discursive antithetical turn between 'city' and 'nature' to suit short-term political–economic imperatives of statecraft.

> Present-day Kolkata, with its lavish, swaggering skyscrapers, swarming population, and ever-increasing, sprawling boundaries, suffers from the dilemma of erasing its past—its deltaic-estuarine-marshy-aqueous origins—while being forced to face this past when pluviometer readings generate alarm and urban utilities and lives get disrupted in a few hours of continuous rainfall, leave aside catastrophic cyclones! (30)

The Adi Ganga–Tolly's Nullah stretch, along with the Kolkata Wetlands, remains to be the only flood relief line for water level resilience in the region as per the study conducted by environmental NGO South Asian Forum for Environment (SAFE), Kolkata, in collaboration with the International Water Management Institute (IWMI), Colombo. Yet 86% of the river/canal stretch flows below the average eflows volume due to excessive pollution and disruption of its natural ecological regime due to contemporary development interventions, adversely affecting the flood-resilience efficacy of Kolkata.

Is the Adi Ganga a river or a canal? What services did it offer to the people and the space? Did these services only pertain to provisional benefits with only tangible outcomes? Does the decay of this ecosystem corroborate with Kolkata's urbanization process? 'One or two unused boats on the clogged and muddy river bank gave some indications of the past', (Mukherjee, 2020: xi) the past preserved in archives in fragments, complemented with real-life accounts or 'living archive' to reminisce the untold story of a tract firmly connected to a city and her socio-environmental resilience.

The king of Oudh, Sagar, who was the 13th ancestor of Lord Rama and the 7th incarnation of Lord Vishnu, performed the Ashvamedha Yajna 99 times. He was desperate to undertake it one more time, but Lord Indra, the king of heaven, who had already performed it a hundred times and had thus earned the title Satamanna, was jealous of being displaced by Sagar. He subsequently stole the horse and concealed it in a subterranean cell, where the sage Kapilmuni was meditating. The 60,000 sons of Sagar started searching for the horse and ultimately they traced it to the place where it was hidden. They assaulted Kapilmuni, believing that he had committed the theft. The sage cursed them and they were burnt to ashes. A grandson of Sagar came to Kapilmuni and begged him to redeem the souls of the dead. Kapilmuni decreed that this would only be possible if the waters of the Ganga (the aqueous form of Vishnu and Lakshmi) could be sprinkled on the ashes. Bhagirath, the great-grandson of Sagar, prayed to Brahma, the Creator, who sent Ganga to earth. Bhagirath led the way as far as Hathiagarh in the 24 Parganas, but could not show the rest of the way. Ganga, in order to make sure of reaching the place, divided herself into numerous channels, and thus formed the delta. One of the channels reached the cell, washed the ashes, and purified the souls which could then reach heaven. Ganga became the sacred stream; the sea took the name 'Sagar'. This junction where the river meets the sea is still worshipped by the Hindus and is a place of Hindu pilgrimage (the legend of Sagar, based on Hunter, 1875: 27–28).

The riverine ecology of the Bengal Delta partially reclaimed to make way to the city of Kolkata (Bhattacharyya, 2018; Mukherjee, 2020) is ingrained in almanac sources – folk songs, myths and mythologies, enabling us to elicit our memories that have not been permanently destroyed but temporarily suppressed under the hubris of mainstream knowledge generated by 'modernity'. The story of King Sagar is a mythological manifestation of geomorphological reality shedding light on numerous streams stemming from the Ganga River in its lower course in Bengal. The present route of the lower Ganges was quite different than its previous route as it flowed through the Bhagirathi channel. The Ganges branched into three streams at Tribeni, near Bandel (Majumdar, 2005), with the Saraswati flowing in a south-westerly direction, past Saptagram, and the Jamuna (not the one in North India or in eastern Bengal) in a south-easterly direction while the Hooghly flowed in the middle, gliding down to Kolkata and then flowing through the Adi Ganga, past Kalighat, Baruipur and Magra to the sea (Map 1). In the 1778 *Bengal Atlas* by James Rennell, the channel was portrayed to be draining into the Bidyadhari River below Baruipur. In the accounts of colonial officials like Hunter (1875) and O'Malley (1914), Adi Ganga has been documented; they pointed out that though this channel had dried up, yet the bed consisted of a series of tanks and the Hindus still considered the stream sacred and cremated the dead on the sides of the tanks excavated in its bed. Even before the British colonizers, Adi Ganga found mention in the sixteenth and seventeenth century cartographic projections by Portuguese cartographers like Jao de Barros and Van Den Brouke (Mukherjee, 1938). The existence and heydays of the Adi Ganga is reflected in the medieval Bengali literature like the Manasamangal by Bipradas Pipilai, the Chandimangal by

Kabikankan Mukundaram Chakrabarti, the Raymangal by Krishnaram Das and other less prominent works like the Satyanarayan Katha by Ayodhya Ram, Shitalamangal by Harideb and Kalu Rayer Geet by Dwija Nityananda. Besides this, the Vaishnavite Saint Shri Chaitanya's journey on the river Adi Ganga is also detailed in the *Chaitanya Bhagabat* thus signifying the importance of this route in the pre-colonial era. Though there is no established theory yet relating to the demise of the Adi Ganga, yet, the drying up of the river is often ascribed to it being artificially linked to the lower channel of the Saraswati River, whereby the latter became the main channel for ocean-going ships and the former became derelict (Reaks, 1919; Bandopadhyay, 1996; Mukherjee, 2016).

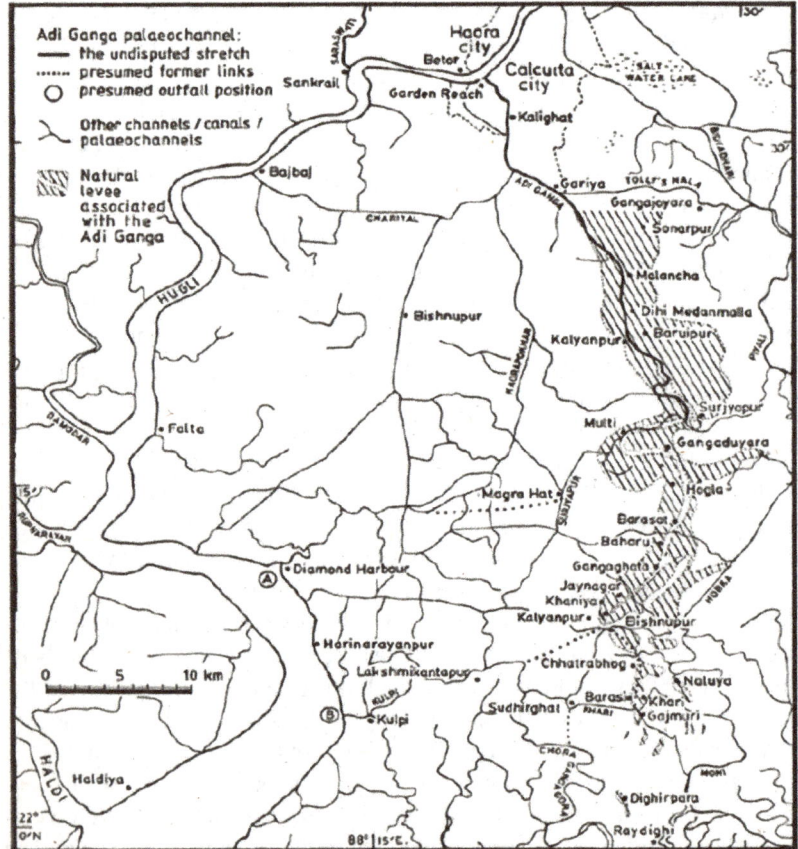

Figure 2.1 The Adi Ganga–Tolly's Nullah route. Source: Bandopadhyay, 1996: 94

2.2.2 *'Calculus of Rule'*

The partial revival of the Adi Ganga and further excavation of the route by Major William Tolly should be contextualized within the larger scheme of excavation of canals and reclamation of marshes that ran parallel to the colonial project of urbanization (Mukherjee, 2015, 2020). On 6 July 1775, Major William Tolly made an application to the government to permit him to excavate the canal between the Hooghly and the Bidyadhari (with the salt water lakes as its spill basin) at his own expense. Out of his initial suggestion where he proposed two alignments, one to the north and the other to the south of Kolkata, the latter got the government's approval along with a temporary land grant, a loan of 100,000 rupees for 12 years and the right to levy tolls at 1% on the price of all goods transported by boats using this canal route. In 1776, he canalized the old bed of the Adi Ganga from its confluence at Hastings, south-eastwards to Garia, a distance of 8 miles and further excavated east to meet the Bidyadhari River at Samukpota, a distance of 9 miles. This entire stretch came to be known as the Tolly's Nullah or the Adi Ganga–Tolly's Nullah after the name of William Tolly. Land deeds and proceedings preserved at the West Bengal State Archives record that the East India Company took over the Canal in 1804. Archival proceedings also state the importance of natives in transaction procedures like collection of tolls – the appointment of Seroo Ghose, as the *suzawal* (tax collector or superintendent) validates the fact.

Historical records also attest the 'constant clashes and negotiations between the government and the corporation over the use of canals as navigation arteries or receptacles of wastewater' including the Tolly's Nullah (Mukherjee, 2020: 64). The Tolly's Nullah like other canals were aggressively maintained for both navigational and urban utility purposes. Various proposals were passed time and again to improve the condition of the Canal facilitating colonial revenue motives. The systematic and detailed documentation of toll collections, net receipts and charges, show good returns on investments which correlates to the frequent and aggressive implementation of maintenance schemes by the Company. Cargo passed through and tolls were collected at three outposts: Russa, Khidirpur and Samukpota (Table 2.1).[1] The traffic on the Canal kept increasing and toll rates were continuously revised to retain the commercial viability of projects. The average toll collections from the canal amounted to 56,829 rupees during 1791–94. Table 2.2 states the annual estimates of tolls levied between 1818–19 and 1824–25, as recorded by the Secretary to the Government Holt Mackenzie (1826).[2]

Buckley's (1883) report stated how despite constant challenges of silt accumulation in the canal requiring costly engineering solutions, the local

Table 2.1 Boat traffic through Tolly's Nullah

Articles	Name of Chowkey	Number of boats	Mundage by canal measurement	Mundage of cargo by estimate
Coal	Samookpota	...	...	...
	Russah	...	...	...
	Kidderpore	203	212,000	123,550
Imported	Do	...	...	...
Fabrics		...	...	...
		4	900	300
Rice	Do	4361	1,363,300	640,975
		3623	102,275	75,725
		132	56,225	29,250
Jute	Do	14	33,050	15,700
		...	...	...
		...	...	...
Indigo	Do	...	...	...
		...	...	...
		5	1150	1023
Salt	Do	...	...	...
		...	...	...
		967	539,100	277,900
Castor Oil	Do	...	...	...
		...	...	...
		2	025	475

Source: Administrative Report on Calcutta and Eastern Canals and Nuddea Rivers, 1868–1869 (Calcutta: Bengal Secretariat Press 1869)

Table 2.2 Collections, charges and net receipts from Tolly's Nullah, 1818–1819 to 1824–1825

Years	Collections	Charges	Net Receipts
1818/19	89,596	7705	81,891
1819/20	88,401	8214	80,187
1820/21	78,495	9621	68,874
1821/22	76,010	7447	68,563
1822/23	67,739	17,477	50,262
1823/24	67,351	7475	59,876
1824/25	68,708	529	62,179

Source: Proceeding 58 dated January 6, 1826 (WBSA, Kolkata).

traffic never faced any decline as it was the only waterway through which 'the rice traffic of the Sundarbans and the surrounding countryside ... was conveyed to the Chetla rice market' (Bandopadhyay, 2018: 64). The proposal to canalize Tolly's Nullah soon gathered momentum as the Port Commissioners of Calcutta wanted to connect the Canal with the wet-docks at Khidirpur where it would act as a feeder to the wet-docks and provide the dock easy access of the boats (Bandopadhyay, 2018; Bhattacharyya, 2018).

2.2.3 *From a Navigable River to a Pillar-Ridden Sewer*

The Tolly's Nullah lost its importance as an artery of navigation and was converted into a receptacle of wastewater. Why and how did the tract lose its viability? The story of this disruption finds 'linear explanations in scientific literature and media reports with the colonial period being projected as the "golden era" and the post-independence period explained as the stage dotted with bureaucratic reluctance' and civic indifference (Mukherjee, 2020: 17). However, applying historical urban political ecological framework, Mukherjee (2020) explains why it is pertinent to contextualize the degradation of the Tolly's Nullah within the overall decay of Kolkata's canal system which started since the late colonial period itself when more profitable technological networks like the railways gradually replaced the former. Using 'archival silence' (with sparser records and reports on the canal trade, investments, etc., from the early twentieth century) as a methodology or even an indicator of decreasing colonial enthusiasm for Kolkata's canals trading outlets, Mukherjee (2020) traces how succession of technologies from inland boat traffic to steam navigation to the railways determined the fate of the city.

From the last decade of the nineteenth century, country boats were cut out by inland steamer navigation 'as means of transport for all the more important articles of merchandise' (Bandopadhyay, 2018: 66). With the increasing momentum of commerce in the Ganges Valley, canal boat traffic in Bengal, connecting Kolkata with its eastern hinterland, 'seemed to be meagre, slow, insecure, and hazardous' (Mukherjee, 2020: 129).

There were constant debates and conflicts of interests among proponents and opponents within the government, among municipal officials and the state regarding the functioning of the canal system as a navigation network or as an outlet serving urban disposal requirements. The proponents of the canal system like O.C. Lees and R.B. Buckley emphasized the significance of canals as feeders to railways, especially during monsoons when huge parts of the delta remained inundated. The then superintending engineer Lees also suggested the canalization of Tolly's Nullah for acting as the link in the steamer navigation and commerce connecting Kolkata with the eastern districts of Bengal and Assam. But the proposal was not approved on the face of excessive competition from the railway lobby, specifically the Eastern Bengal Railways. Lees lamented,

We have been so intent upon obtaining a direct and immediate return upon our capital outlay that we have nearly killed our golden goose and have scarcely given a thought to the indirect benefits likely to be conferred by good waterways and cheap transport.

(quoted in Bandopadhyay, 2018: 96)

Bandopadhyay (2018: 97) adds, 'The "golden" goose is now almost dead. Putrid smell it now exudes is not only prejudicial to public health and hygiene but also a disgrace to municipal administration'.

Raw sewage and solid waste on its banks were dumped into the Tolly's Nullah. A large number of sewerage drains belonging to the Kolkata Metropolitan Corporation (KMC) and Kolkata Metropolitan Water and Sanitation Authority (KMWSA) kept on discharging directly into Tolly's Nullah on both of the banks. With these outlets being ungated (with no lock gates constructed on them to check and regulate the flow of water during high ebb-tides) the pollution level of the Tolly's Nullah intensified with an alarming rate. A study conducted by an environmental NGO, Vasundhara Foundation, pointed out that the increasing pollution load was a major hindrance as the insufficient cross-sections of the canal did not allow for the flushing of pollutants by the tidal waters.

In regards to the urgency of the situation, CEMSAP (Calcutta Environment Management Strategy Action Plan) presented its report in 1997 that provided detailed insights for the management of EKW and the canal system (CEMSAP, 1997). It emphasized on the revitalization of Tolly's Nullah by connecting it with the Piali River and through the construction of lock gates that would control and regulate the flow and discharge of wastewater. According to a report published in the Bengali daily Ananda Bazar Patrika (21 December 2000), the Urban Development Department, GoWB, had plans for renovation of the 15.5 km stretch of the canal between Hastings and Garia. However, the implementation of the metro railway extension (from Tollygunge to Garia) over the canal, which almost dissected the river, threw a wrench on these plans. Three hundred pillars were placed into the canal bed, each at a distance of 20 m from the next (Mukherjee, 2016). The metro rail extension project was approved despite protests, petitions and

Figure 2.2 The Pillar-ridden Tolly's Nullah. Source: Author (Mukherjee)

litigations from various sections of society. The railway authorities sought no environmental clearance and the project was sanctioned in spite of violation of the Environmental Protection Act (EPA), 1986, and Environmental Impact Assessment (EIA) Notification, 1994. The state used the archaic Section 11 of the Railways Act, 1989, a remnant of the colonial revised edition of the act of 1890 that gave railways access to construct 'upon, across, under or over any land, any rivers, canals, brooks, streams or other waters', thus implementing the project with tremendous socio-ecological cost. 'The outcome was the transformation of our heritage river into a pillar-ridden sewer' (Mukherjee, 2010: 138) (Figure 1).

2.2.4 Shades of Environmental Activism

The geo-cultural attributes of the 'river turned nullah' took a new turn in the 1990s with an uproar over its restoration. The forced defunction of Adi Ganga and the subsequent eviction of the squatter population was reflective of the state's agenda of enforcing the corporate perspective of development that promoted urban growth at multiple environmental costs. Kolkata visualized various waves of protests with differentiated agendas overtly manifesting conflicting ideals, yet converging together to resist authoritarian developmentalist visions and interventions with severe environmental and social costs.

The environmentalism surrounding the river was also segregated among different actors with different motives that influenced the activism's narrative as well as its outcome. The movement commenced with the conservationist approach in the 1990s under the aegis of Rebati Ranjan Bhattacharya who rallied up NGOs for almost a decade to pressurize state agencies regarding the clean-up of the canal and its banks, which had been transformed into a stagnant pool of liquid waste. This affected the local urban upper- and middle-class sensibilities of people who had grown up on the banks and often reminisced about the once-functioning river. *A Plan for Adi Ganga–Tolly's Nullah*, a rare document which was developed by the Tollygunge Development Council (TDC), a city-based environmental NGO (accessed from the primary source collection of Mohit Ray's personal library), showed TDC's attempts to gain attention of I&WD and KMDA officials regarding the cleaning up of the river banks as early as March 1981. Subsequently, in April 1982, TDC submitted *Resuscitation of Adi Ganga and Its Merger with Re-excavated Nischintapur-Hotor Khal for Navigation, Pollution Control and Small Irrigation Facility*, a comprehensive action plan for the river's restoration, to the government. In later years, several plans were formed between the NGOs and state agencies to revitalize the Nullah, although none of them materialized (Mukherjee, 2020).

Bhattacharya's own conservationist approach of resuscitating and regenerating the Nullah as part of the heritage river, the Adi Ganga, came from his bond to the waterbody and witnessing its rapid transformation from a navigable canal to a Nullah, and from a heritage river to a contaminated and foul sewer line. A writ petition filed by Bhattacharya in 1996, forced the Green Bench of Kolkata High Court to hear out the issue, followed by the High Court order on 2 February 1997, which ordered the state to rejuvenate the channel by cleaning it and clearing out the squatter settlements on both sides of the banks (Mukherjee, 2020). This led to the formation of a committee, headed by Chief Secretary of State and consisting of Chief Engineer, KMDA, the Commissioner, KMC and the Secretary, I&WD. INR 40 crore was recommended by the committee for the Tolly's Nullah restoration project in May 1997, but this venture proved futile after R.P. Sharma, the financial adviser of National River Conservation Directorate (NRCD) visited Kolkata to meet with the representatives of KMC, KMDA and I&WD, demanding a detailed report on the present state of the Adi Ganga River, on the basis of which the grant would be approved. However, no action or steps were taken by the state authorities in this matter and, in 1998, according to a revised budget submitted to NRCD, INR 29 crore and 50 lakhs were sanctioned for the project, but the canal squatter residents demand for rehabilitation scheme, mobilized by grassroots organizations, halted the project's progression. The residents, belonging to lower economic classes were part of the metropolis' informal workforce, had nowhere to go and therefore wanted to be rehabilitated in the pre-implementation phase of the project. However, as time progressed neither the rehabilitation plan nor the project came into force.

The publication of CEMSAP report in 1997, not only established the link between the EKW and the canal systems of Kolkata, but also suggested the integration of ecotourism and economic needs through the restoration of the canal, that would facilitate trans-shipment of commodities and commuters. The experts also proposed a plan to connect the eastern extension of Tolly's Nullah with Circular Canal, Bhangarkata Canal and Krishnapur Canal by excavating another 11 km of a new water stretch that would give the decaying river a new lease on life, but all talks halted in 1999 with the metro rail extension plan. Rebati Bhattacharya, along with other leading environmentalists of the city such as Sankha Ghosh, Gautam Bhadra, Subhash Dutta and Mohit Ray, concerned with how the project would infringe upon the canal's natural flow, became actively involved in gathering information about the project and submitted a letter to the metro rail authorities regarding the same. In order to pacify the emerging uproar, the metro rail authorities responded that all decisions regarding this project would commence after consulting with the I&WD and keeping in mind the concerns for the river Adi Ganga. The falsity of this assurance became apparent when the plan details unfurled, revealing the plans to concretise the waterbody with 300 pillars of 2 m diameter each, placed at 20 m intervals across the Nullah, with 5 m headroom over the decks of existing bridges on the river (Mukherjee, 2010, 2016). This plan was a testament to the state's ignorance of the river's cultural heritage and the ecological importance to the city itself.

The realization that this will be the end of the Adi Ganga, set in motion the chapter of Kolkata's own river activism that witnessed various environmental groups, intellectuals and activists unitedly denouncing the state's technocratic and ecologically ignorant decisions. The first protest meeting was organized on 25 April 2001 at the CSSS auditorium by the environmental NGO Vasundhara Foundation that was also attended by several other NGOs such as Disha, TDC and Calcutta 36 and leading intellectuals, activists and academics. Mohit Ray, the secretary and scientist of Vasundhara Foundation, explained in this meeting the unscientific and unethical course of the project, citing not only environmental but legal concerns as this extension project violated the EIA Act guidelines. The other attendees discussed the rich history of the megapolis and the Tolly's Nullah, while also engaging in the present issues of degradation and encroachments. This issue took a legal course when Subhash Dutta, a renowned environmental activist of Kolkata, filed a writ petition against the metro extension project in Kolkata High Court in January 2001. However, the case was lost after the court ruled in favour of the metro rail project, citing the archaic Section 11 of Railways Act 1989, a continuation of the Colonial Act of 1890 that envisioned ecological resources from colonies as nothing but raw materials to exploit and sadly gave the railways leeway to utilize any ecological resource for their expansion. The state's manipulation of the legal discourse, however, failed to dampen the environmentalism, as city activists cited the Environmental Protection Act of 1986, which was drafted as a commitment to the UN's 1972 Stockholm Conference, and asked it to be given preference for conservation of an ecological resource. They also organized another major meeting in April 2001, consisting of representatives from various pro-conservationist networks such as Calcutta 36, Nagarik Mancha, Utsa Manush, Purba Kolkata Paribesh Sameekshan, Disha, Swastha-O-Paribesh, Sahay, Bigyan O Bigyankarmi, TDC and Kaladhawani. However, the metro rail authorities started their project work based on the High Court ruling and unfortunately this lessened the uproar for the river's conservation as well as any restructuring schemes.

Although the heritage river campaign had stoked waves of protests across the city's intellectuals, it was another mobilization effort that had emerged simultaneously in the region that didn't include the environmentally aware middle class as participants. This movement, spearheaded by grassroots organizations, focused on advocating for the survival and sustenance rights of the dwellers of the canal banks whose livelihood opportunities and shelter were threatened by the metro extension project as well as the river rejuvenation activism. These canal bank residents were squatters, consisting mostly of marginalized migrants who had settled in Kolkata after partition, mostly from Sundarbans in search of employment opportunities. As victims of landlessness from development and natural disasters and the unending cycle of poverty, the migrants often had to leave their own homes and travel to the urban regions in search of some form of employment and shelter, which for them have always been railway tracks, pavements, under the bridges and canal banks. The CEMSAP report states that in the 1990s the Tolly's Nullah stretch was habituated by approximately 41,500 people who were employed in the informal urban sector, with males working as rickshaw pullers, small-time vendors, construction workers, and females as maids, nursing attendants and construction workers (see Table 2.3).

The squatter population, whose multifaceted contributions in the form of informal labour, were instrumental in the city's corporate beautification and development projects, and their own shelter was threatened under the state's narrative of growth and progress. While the official records state that the encroachers peacefully vacated their shelters when notified, the original testimonials are far from the claim. This authoritarian perspective of modernization at the cost of the canal bank dwellers sparked protests across the margins which unfolded through the subaltern resistance against the state–corporate nexus, spearheaded by the Ucched Bachao Jukta Mancha (UBJM), a radical grassroots organization, determined to deter the eviction and demolition process by standing in front of the bulldozers (Hindustan Times, 2001). They were in the forefront of the movement in September 2001 with various other organizations such as Manthan, Association for Protection of Democratic Rights and the support of intellectuals such as Badal Sarkar, Nabanita Deb Sen, Gautam Chattopadhyay, Debesh Roy and others (Statesman, 2001). The protest gained a political colour with

Table 2.3 Squatter population on the banks of Tolly's Nullah

Stretch	Families	Number of people
Hastings to Tolly Golf Club	2,500	12,500
Tollygunge to Garia	1,300	6,500
Garia to Samukpota	450	2,250
Boat Canal to Tolly's Nullah	4,350	20,250

Source: CEMSAP (1997: 140).

the opposition parties such as Trinamul Congress Party (TMC) and left front partners like Revolutionary Socialist Party and Forward Bloc. On 11 September 2001, the day of the eviction witnessed violent clashes at Garia Bridge, between the protestors from UBJM and TMC with the Rapid Action Force who continued the demolition drive and openly used lathi charges, not sparing women, children or the elderly (Times of India, 2001c). The brutal attack on the protestors who were fighting for the ones most closely associated with river drew the ire of various organizations. Feminist organizations like Nari Nirjatan Pratirodh Mancha, All India Progressive Women's Association, Sanlap and Maitree highlighted the plight of women and how they were the worst victims of the eviction processes with their loss of security. Among the most unfortunate sufferers of this eviction drive were the squatter children who were left unprotected with no shelter and an uncertain future. In response to this state brutality, various children submitted a mass petition to the Juvenile Justice Board to protect the children affected by this project. On United Nations Human Rights Day, i.e., 10 December 2001, UJBM organized a sit-in demonstration by children outside the Taj Bengal, a five star hotel in Kolkata, in front of the international planners at Cities Alliance, to showcase the plight of the marginalized communities and the ignorance for their well-being by the state. Although the UJBM continued to rally protests and create community kitchens to feed the squatters, the eviction drive continued, with police brutality eventually finalizing the evacuation procedure (Mukherjee, 2020).

While two varieties of environmental activism emerged and conflicted with each other over the conservation of Adi Ganga River, none of these successfully prevented the death of the river or the squatter eviction from its banks. The reason behind this can be encountered in their diverse types and patterns of protest strategies and mechanisms. While environmental activism, which only advocated river conservation, took the course of non-violent methods such as media activism, seminars, peaceful rallies and judicial activism, the grassroots activism often took a brutal turn, as evident through violent clashes between police, activists and victims who were participating in the protest movement. The difference between the two campaigns and their lack of integration in recognizing the common cause played an important role in the outcome of the movement. On 26 December 2000, a group of 'eminent Kolkatans' rallied for a few kilometres across the stretch of the Ganges to highlight their concern for the river, but this Walk for River rally, which meant to lend support to the new state initiative by government agencies, business groups, international bodies and NGOs 'to develop the waterfront for this city and also to raise awareness about a clean river', interestingly ended at the Millennium Park on the Hooghly Riverfront, developed and inaugurated by the KMDA in 1999 as part of the river beautification scheme (Ray, 2001a). The radical protest groups on the other hand joined by (metro) project-affected marginalized shanty dwellers, claimed that the fight for Adi Ganga was a fight for their existence, and stated that 'This was Kolkata's very own Narmada. The ingredients were there all right, albeit on a much smaller scale' (Times of India, 2001c). Environmental activists like Mohit Ray was one of the few heritage activists who spoke against displacement of 'encroachers', countering the initial argument presented by the metro and state authorities that one of the reasons for constructing the pillars on the river was not to displace the squatters, but only to bring into fore the state legitimization of hugely funded 'development' prospects and their indifference to the ecological perspectives.

The heritage argument was strongly imbued with upper- and middle-class awareness of environmental activism and historical and cultural sensibilities, with the use of vernacular and colonial archives linking the city's rich historical and religious past with the heritage river. Examples from the West, especially the American Heritage River Initiative, that revitalized community rivers and their banks to celebrate their heritage, was cited to showcase the narrow-minded approach of NRCD towards river preservation that ignored the environmental, economic, cultural and historic value of old rivers in Third World cities (Ray, 2000; Hossein & Ray, 2000). Apart from heritage, there was a lot of nostalgia surrounding the river among the locals who had grown up beside it and developed a bond with the waterbody, which fuelled this fragment of activism. On the other hand, the grassroots activism of the squatters was their fight to coexist, through basic needs and rights, which were dehumanized in favour of the river's protection. While the squatters were blamed by the environmental activists for the degradation of the river, and their presence was cited as a series of potential health and ecological hazards, according to Sushovan Dhar, activist and member of Manthan, the grassroots resistance reflected the squatters' strong will to maintain the canals in future with their own toil only in exchange for shelter on the banks (Mukherjee, 2020).

2.3   Conclusion

A short walk on the present stretch of the Adi Ganga, also known as Tolly's Nullah, would reveal its dilapidated state, permeated with filth, waste and cemented proofs of 'progress' and occasional squatter residences. This comes in stark contrast with the memories of the heritage river that was once the main flow of River Hooghly and a major distributary of the Ganges river in the Bengal Delta. What was worshipped and revered not only for its geo-cultural attributes but also for its transactional benefit, leading to the rise of Kolkata's water transportation and economy, today lies defunct and forgotten. The legacy of the river received the final nail in the coffin with the state legitimization of corporate agenda trumping over sustainable solutions for development and the multiple socio-ecological conflicts over the river banks access. While various shades of protests highlighted the different motives and objectives by the differing groups calling attention to the safety of the heritage river, the metro extension project derailed any modicum of hope for the river's survival. Before delving into the modalities of river protection schemes and state-activist group conflicts in the curious case of Adi Ganga, it is necessary to consider the importance of the 'heritage' narrative practised by various groups for the river's conservation.

The case of Adi Ganga critically interrogates the whole idea and definition about heritage as well as its stakeholders. Whose river is this? Is it of the people who have learned of its stories from various ancient native literatures, or whose ancestors have grown up on its banks, or of the marginalized communities whom it gave shelter to in their most trying times since the partition days and whose livelihoods depended on its resources? It is necessary to consider the fact that if the heritage of Adi Ganga encompasses its attributes across a spatio-temporal scale, then it must also be inclusive of the diversity of its participants, actors engaging in the river's heritage activism. But is it possible to explore the heritage discourse surrounding the Adi Ganga by a spatial mapping of the socio-political conflicts focusing on the differing agendas, motives and affiliations? The Adi Ganga conundrum is one such example that highlights the importance of a concentrated analysis of local issues and their part in the environmental dynamics of a region.

Notes

1 Chowkey in the local dialect means toll collection centre.
2 Proceeding 58 dated 6 January 1826 (WBSA, Kolkata).

References

Bandopadhyay, H. (2018). *History of canals in Bengal*. Kolkata: Doshor Publication.
Bandopadhyay, S. (1996). Location of the Adi Ganga paleochannel, South 24 Parganas, West Bengal: A review. *Geographical Review of India, 58*(2), 93–109.
Bhattacharyya, D. (2018). *Empire and ecology in the Bengal Delta: The making of Calcutta*. Cambridge: Cambridge University Press.
Black, R. (1990). 'Regional political ecology' in theory and practice: A case from Northern Portugal. *Royal Geographical Society, 15*(1), 35–47.
Blaikie, P. (1985). *The political economy of soil erosion in developing countries*. London: Routledge.

Blaikie, P., & Brookfield, H. (Eds.). (1987). *Land degradation and society*. London: Methuen.

CEMSAP (Calcutta Environment Management Strategy Action Plan). (1997). *Management of East Calcutta wetlands and canal systems: A report*. Calcutta: Department of Environment, Government of West Bengal.

Harrison, R. (2013). *Heritage: Critical approaches*. New York and Oxon: Routledge.

Hossein, M., & Ray, M. (2000). *Oitijhyer nadi A. Gangake banchiyei metro railke cholte hobe*. B-Sambad.

Hindustan Times. (2001, September 22). Squatters vow to resist dawn strike.

Hunter, W. (1875). *A statistical account of Bengal: 24 Parganas, vol. I: Districts of the 24 Parganas and Sundarbans*. London: Trubner & Co.

Majumdar, R. C. (2005). *History of Ancient Bengal, 1971*. Calcutta: Tulshi Prakashani.

McCarthy, J. (2002). First world political ecology; lessons from the wise use movement. *Environment and Planning: Part A, 34*(7), 1281–1302.

Mukherjee, J. (2010). *Development and planning: A social biography of the canals and wetlands in Kolkata* [Unpublished dissertation]. Kolkata: Jadavpur University.

Mukherjee, J. (2016). The Adi Ganga: A forgotten river in Bengal. *Economic and political Weekly, 51*(8).

Mukherjee, J. (2020). *Blue infrastructures of Kolkata: Natural history, political ecology and urban development in Kolkata*. Singapore: Springer.

Mukherjee, R. (1938). *The changing face of Bengal*. Calcutta: University of Calcutta.

O'Malley, L. S. S. (1914). *Bengal district gazetteers: 24 Parganas*. Calcutta: Bengal Secretariat Book Depot.

Ray, M. (2000, April 17). Review. *Statesman*.

Ray, M. (2001a, January 5).On the waterfront. *Statesman*.

Ray, M. (2001b, March 3). Metro runs over law in southern rush. *Hindustan Times*.

Reaks, H. G. (1919). Report on the physical and hydraulic characteristics of the delta. In *Stevenson Moore committee report on the Hooghly River and its headwaters* (Vol. 1). Bengal: Bengal Secretariat Book Depot.

Robbins, P. (2002). Obstacles to a first world political ecology: Looking near without looking up. *Environment and Planning: Part A, 34*(8), 1509–1513.

Statesman. (2001, September 21). Confusion clogs Tolly's Nullah.

Times of India. (2001a, September 23). Machine raze futile rage.

Times of India. (2001b, December 11). Children protest eviction.

Times of India. (2001c, September 30). Muddy flows the nullah.

Walker, A. P. (2003). Reconsidering 'regional' political ecologies: Toward a political ecology of the rural American West. *Progress in Human Geography, 27*(1), 7–24.

3 The Opportunities and Challenges of Transboundary Conservation – Solutions in Adaptive Management

Radhika Bhargava

3.1 Introduction

The boundary between land and water, nature and society, two states and two nations are the borders that have been impacting how nature has been conserved for a long time. In some areas, the border between nature and society is given extreme importance, and the conservation of nature is entirely ignorant of the needs of society. Where, not only the needs of the people are ignored, but also nature's need of the people is ignored. The result is a socially failed conservation approach. The debate on the importance of the border between nature and society is an unresolved and ongoing one. But, the discussion of this border becomes even more pronounced when there is another border that politically divides the region of interest into one or more territories. The conservation of this system calls for a transboundary conservation approach. The distinctive feature that sets apart transboundary conservation from other regimes of conservation governance is the essence and presence of a 'border'. This chapter focuses on understanding the border in the context of transboundary conservation of natural resources. The two questions it answers are: (1) how does a border necessitate transboundary conservation? and (2) under what circumstances can transboundary conservation be effective? Let us begin with understanding the basics of transboundary conservation.

Transboundary conservation is an approach taken to govern and manage natural resources shared across the different political, legal, institution, social, cultural or economic regimes for conservation. The practice of designating areas for conservation has been around since the 1800s, and since then the practice of designating areas across politically defined borders has existed. However, a clear definition of transboundary conservation did not come into place until the late 1900s. Today, there are more than 200 conservation sites that encompass political borders.

DOI: 10.4324/9780367486433-3

The understanding, definition and demarcation of a border is a central theme to the transboundary conservation approach. Based on the currently managed transboundary resources, IUCN has defined three types of borders that are important for natural resource management (Sandwith et al., 2001). The first one is the boundary between and within states. The boundaries between nations have the usual connotation of the international boundaries between two or more countries, whereas the border within a nation holds importance for those nations where the constituent states or provinces also have the autonomy of governance of their natural resources. While the concept of political borders in terrestrial systems is relatively easier to understand when the concept of political borders is assigned to marine ecosystems, the definition of the border becomes 'fluid'. Still, the need for connectivity in this fluid space becomes ever so important. Thus, marine borders make the second category of borders that the transboundary conservation approach works with. The last category, as defined by IUCN, is of abstract notions where border partners can include nations or states that are not adjacent to each other but are directly involved with the management of natural resources. This category also includes different government/non-government players as stakeholders of the commons.

3.1.1 Approaches to Transboundary Conservation

There are two leading organizations – International Union for Conservation of Nature (IUCN) and the World Commission on Protected Areas (WCPA) that have extensively worked towards promoting transboundary cooperation internationally. They have divided international collaboration on the management of shared resources into four different types of agreements. The different types are: a transboundary protected area, transboundary conservation landscape or seascape and transboundary migration conservation area, as well as a park for peace as a special designation.

A transboundary protected area is defined as an ecological area comprising multiple protected areas shared between one or more nations with varying degrees of cooperation across the international border. A transboundary conservation landscape or seascape is different from a transboundary protected area. It is an amalgamation of protected and multi-use areas that sustains both human and ecological processes and is shared between one or more nations across international boundaries. If an area is completely protected and is shared across political borders, then it is of the transboundary protected area category. But, if an area also supports multiple resource use in addition to conservation, then it comes under the second category. A transboundary migratory conservation area or migratory corridor, on the other hand, only focuses on those wildlife habitats across two or more countries that are crucial to sustaining the populations of migratory species, again with varying degrees of international coopera-tion. In this case, the transboundary partners do not have to be geographi-cally adjacent. Lastly, the special designation of Peace for Parks is given to any of the above-defined areas when they also promote international peace (Vasilijević et al., 2015).

Examples of these different types of transboundary conservation areas can be seen from all around the world. Some of the iconic ecosystems that are protected as Transboundary Protected Areas (TBPAs) are the Victoria Falls in Zimbabwe and Zambia and the Iguazu Falls in Brazil and Argentina. Even the only gorilla habitat that is shared between Rwanda, Uganda and Congo is a TBPA. The Pantanal wetland landscape is an exam-ple of a transboundary conservation landscape from South America where Bolivia, Brazil and Paraguay protects the world's largest tropical wetland. This landscape stretches across about 40 million acres supporting thousands of plants, animals, including migratory birds in addition to supporting mil-lions of people that depend on the landscape for flood protection and water resources. A spectacular example of a migratory corridor transboundary landscape is in the Ombai Strait shared between Indonesia and Timor Leste in the Lesser Sunda Ecoregion. This region forms an important migratory corridor for cetaceans like the sperm whale, blue whale and whale shark. An example of Peace for Parks came from 1932 when the USA and Canada established Waterton-Glacier International Peace Parks to celebrate their long-standing peaceful relations.

These sub-types of transboundary conservation are based on the purpose of formation of the area. Some of these areas can also be established to meet multiple goals from different sub-types. The purpose of the formation of a transboundary conservation area is also dependent on its governance. Different governance mechanisms can be adapted to address the needs of the transboundary conservation sites. These governance mechanisms rely on the political framework of the different nations/parties involved in trans-boundary conservation.

3.2 Transboundary Conservation Governance

The implementation of transboundary conservation exists at the heart of environmental governance. Starting from the formation of such an area to meeting the purpose of the formation, it involves one or more nation-states, their jurisdiction and power arrangements. Graham et al. (2003) have defined transboundary governance as – 'the interactions among structures, processes and traditions that determine how power and responsibilities are exercised, how decision are taken, and how citizens or other stakeholders have their say'. Here the interactions are among one or more jurisdictions, and institutions across one or more nations as the territory of interest goes beyond these legal and geographical structures. The need for transboundary governance is formed by the interdependence of these nations to meet a desired goal/purpose for this shared territory. As is, none of the involved nations can fulfil this purpose individually. Thus, transboundary governance involves multiple nations, but the scale and level of involvement vary across different scenarios.

Different types of governance structures can be adopted for transboundary governance. The governance can be top-down where the highest political authority at the federal level with their national/sub-national ministries oversee the management, and they directly oversee the implementation of transboundary conservation. When multiple federal/national agencies oversee the implementation, then the governance takes the form of shared governance where various actors and institutions work together. In some cases, the autonomy of implementation is given to private landowners, NGOs, academic institutions or for-profit organizations forming a privatized governance framework. Lastly, governance can involve local communities and indigenous people that directly depend on the geographic area in the form of a bottom-up approach to governance.

3.3 How Does a Border Necessitate the Need for Transboundary Conservation?

The establishment of a protected area for conservation is a complex task. An added layer to this complexity arises when there is a border diving the land into different political/ecological/institutional units. Often the border makes it harder to establish a region for conservation (Westing, 1998). But it also gives an added advantage (for example promotion of peaceful measures) to establish borderland areas as conservation priority sites to meet social, environmental and political objectives. It has been established in conservation science that the scale of the conservation site matters in meeting conservation outcomes (Dudley et al., 2014). The region of interest should be big enough to sustain the significant ecosystem and social functions that address the need for the establishment of the transboundary site. The first purpose is that it can meet the scale requirement by connecting critical ecological zones that are distributed across the border. Secondly, because of establishing a big conservation site across the border, essential wildlife corridors can be linked. And lastly, the cross-border relations can be strengthened among the states, and the lives of the people divided by the border can be linked. Although the ulterior purpose of establishing a transboundary conservation area is the preservation of biodiversity, the underling opportunities can be varied and tied around the significance and role of the border to the region. There are various opportunities provided by the presence of the border that can be achieved by forming a transboundary conservation site. Through these potential opportunities discussed below, the presence of border in biodiverse regions necessitates the formation of a transboundary conservation site.

3.3.1 Opportunities from Transboundary Conservation

The establishment of a transboundary conservation area is aimed towards achieving several multifaceted and interdisciplinary benefits. Unlike the strict-use protected areas that are usually only established to generate ecologically focused benefits (West et al., 2006), the benefits associated with transboundary conservation sites usually cover ecological, social, economic, cultural and political benefits. Researchers are, however, looking at how useful are transboundary conservation sites in providing these benefits on-the-grounds. Some of the crucial arguments regarding the effectiveness of transboundary conservation approaches from around the world would be discussed later in this chapter. For now, let us understand the different categories of opportunities that the transboundary conservation approaches can provide, which gives buy-in for their establishment.

3.3.2 Ecological Opportunities

First and foremost, the overarching purpose for the establishment of a transboundary conservation site is ecologically motivated. Transboundary conservation sites are often necessitated because the individual protected areas across the borders are not sufficient to meet essential ecological functions to sustain conservation of a habitat or species. Individual and unconnected conservation sites are often not capable of providing the connectivity needed for wildlife migration or of maintaining a significant genetic pool. Or, in some cases, the external pressures from one side of the border also impact the natural resources on the other side that the inclusion and protection from across the border become necessary.

In this way, forming a transboundary conservation unit brings in ecological benefits by improving the connectivity of the region of interest. By improving the connectivity of the protected sites, the long-term viability of target species, like the Grizzly Bears, Royal Bengal Tigers, Polar Bears, etc. can be improved. If these sites were to be managed as separate units than the smaller and individual habitats would not be able to support a larger population as compared to a well-connected cross-border site can at a large scale. Even in the case where transboundary areas are not adjacent, like in the case of a transboundary migratory corridor, migrating species can be ensured of healthy and protected habitats throughout their routes of migration. One of the issues with isolated protected areas is that critically endangered species are often present in small patches (Mbora & McPeek, 2010). Due to excessive pressure from outside of the protected sites, the target species is not able to grow in numbers. The pressure keeps on pushing them into smaller populations. Although protected, such smaller populations are not able to survive. When the habitats are protected at a larger scale, such isolated species can grow in number and area, which helps in maintaining their significant numbers. This can reduce the fragmentation of ecosystems and promote integrated systems that can support various ecosystem processes for a healthy habitat. A healthy ecosystem is protected against the impacts of natural disasters like a forest fire or a cyclone. These protected habitats also have limited impact on the system as compared to when the system is fragmented where the losses can be much higher to the point that an ecosystem can be obliterated as a result of a single extreme event (Casagrandi & Gatto, 1999).

An integrated, well-connected and well-managed transboundary conservation area has the potential of giving significant ecological benefits. These benefits range from smaller-scale benefits to a species or large-scale global benefits like supporting global biodiversity regimes or regulating global temperatures. One of the key benefits of such an integrated system is providing resilience in the face of climate change. Achieving climate change resilience needs a health system that can adapt, evolve and sustain during the impacts of climate change. One of these adaptations could be migration to a more suitable habitat (Jennerjahn et al., 2017). Resilience can be achieved in a well-protected and well-managed ecosystem that supports the hierarchy of the food web and different ecological processes (Araújo et al., 2004). This scenario can be exemplified through the Northwest Forest Plan, which is one of the first forest management plans that was based on the concepts of regional conservation planning (Ferrier, 2002). Different ranges of focal species and endemic species were included in its conservation planning that spanned large areas. When the current conservation plan of this region was analyzed against the effects of climate change, it was found that the currently protected ranges will lose their integrity as species migrate to more climate-wise suitable but non-protected areas. This could lead to the extinction of key species, especially the Northern Spotted Owl, which is an umbrella species of the region (Carroll et al., 2010). Losing an umbrella species could ultimately lead to the disintegration of the ecosystem. Thus, in the region with large areas of natural forest cover, reserves should be designed to be more resilient to climate change to ensure that a significant component of diversity is preserved.

In this way, transboundary conservation areas when they meet the characteristics of being large scale, well connected, inclusive and integrated can provide several ecological benefits. These ecological benefits play a crucial role in maintaining local and endemic species but also supporting global ecological processes. The critical piece here is if and only if these transboundary conservation ecosystems are well-managed can they provide the desired ecological benefits.

3.4 Socioeconomic Opportunities

Prior to the partition of the divided landscapes, people would have been able to move and build lives around the region. But after political separation, their lives would have been divided. The cultural, social and economic ties that were present before the formation of the border does not have to suffer through the partition. The impact of borders on the lives of the borderland people is worsened when their dependence on the natural resources goes beyond the political notions of space, and become worst when the borderland population is marginalized (Salisbury et al., 2011). In these cases, if a transboundary conservation area is formed to give not only ecological but also social benefits, then the society can be equally benefited from conservation initiatives. As explained by Christie (2004), a socially ignorant conservation plan that only focuses on the needs of the natural system is often successful only on papers and is a massive social failure. Thus, transboundary conservation managers must design socially and ecologically sound conservation plans. If, and when these plans are socially inclusive, several social and economic benefits can be generated like the promotion of trade, tourism and culture, and resolution of cross-border human–wildlife and armed conflicts.

When a border separates two regions, it also hinders the movement of people and goods. When goods cannot move across the border, or at least not freely, then it causes a shortage of supply and income on both sides. The transfrontier territories work with the local suppliers and cross-border buyers to smoothen the movement of goods. Promoting trade across a border is often the leading reason for the establishment of parks as the goods in question are often nature-based goods that depend directly on the health of the ecosystem. The transboundary conservation network between Lesotho and South Africa exemplifies the purpose of establishing a trade and commerce network through the establishment of the Maloti-Drakensberg protected area network (Vasilijević et al., 2015). The resource in question for trade is dependent on the ecosystem service provided by the natural resources. A well-connected trade network is necessary for a win-win situation for both Lesotho and South Africa. The resource in question was the river that flows in the area and the number of regulatory, provisional and economic services provided by it. The transboundary conservation area supported the sustainable management of the river so that hydrological services can be received by South Africa while maintaining a healthy ecosystem for livelihoods in Lesotho. This was achieved by establishing payment for ecosystem services, a process which monetizes the ecosystem services when used by the user, and the money is then given to support the livelihoods that depend on sustainability maintaining the system.

An extension of uniform trade networks across a border is the formation of a cross-border tourism industry. Tourism in conservation areas is used as a tool for creating awareness about conservation, generating livelihood options, securing funds for conservation and promotion of the protected reserve. Tourism in conservation sites should always be sustainable, community-based and eco-friendly; otherwise, all the hopeful benefits from the industry are nullified. Tourism, when performed in conjugation with conservation, provides significant socioeconomic benefits. The tourism industry creates several jobs which have the potential to support and uplift the local livelihoods. They also help in the promotion of conservation awareness, local culture and tradition, and funding of other conservation activities. One of the most useful implementations of a transboundary ecotourism industry can be seen in Africa between Angola, Botswana, Namibia, Zambia and Zimbabwe under the name of Kavango Zambezi (KAZA) and in Europe between Italy and France (Vasilijević et al., 2015). Italy and France have been working together since 1987 to co-manage the transboundary Maritime Alps Regional Park situated in Italy and the Mercantour National Park in France for the preservation of their key species. The conservation was so successful that the region has developed, implemented and completed multiple improvised plans since its establishment. Some of the successful plans include an ecotourism network with well-connected hiking trails, multilingual signage and a car-free tourism infrastructure that goes across Italy and France. These plans also involved an extensive network of local stakeholders who are actively participating with tourism, education and management of the park. This network park now stands as a stronghold promoting regional biodiversity, history and culture. Similarly, in the Kavango Zambezi (KAZA) transboundary park with a combination of conservation and tourism, the park promotes the joint culture, livelihood and natural resources while ensuring the unrestrained movement of both the wildlife and the humans in the park.

Transboundary conservation areas are rife with the essential cultural, spiritual and traditional context that makes these sites significant not only from a historical perspective but also from the perspective of a living heritage. The values are upheld of utmost importance by the local indigenous and tribal groups, and by other populations who have sacred values associated with the natural resources. Another important aspect is the identity and representation people have with these regions.

Just as important it is to make conservation plans socially relevant, it is even more critical to make sure that the rights and needs of the local indigenous and tribal groups are central to the transboundary area (Chan et al., 2012). Thus, a well-designed transboundary conservation site can support the local indigenous and tribal populations from across the border whose social structure do not necessarily conform with the political definitions of a border. Transboundary conservation approach can also support the local traditional knowledge that has been around for ages. And that has been successful in supporting the natural and cultural regimes while maintaining peaceful relations among the different tribes and indigenous groups that belong to the area. These cultural benefits of transboundary conservation sites are incorporated in the tri-nation Mount Kailash Sacred Landscape encompassing India, China and Nepal in the Hindu Kush Himalayas. The region is spread across 31,000 km^2 and supports about one million people of various indigenous groups and tribes from the three countries. Mount Kailash regional landscape is identified important for its immense cultural, biodiversity and social values. In terms of the cultural aspects, it is sacred for five religions (Hinduism, Buddhism, Bonism, Jainism and Sikhism) that are not only active in the region but are active worldwide and hold sacred sentiments towards Mount Kailash and Lake Mansarovar (Sayer et al., 2013).

The collaboration between different nations for the formation of transboundary conservation areas is a sign of these nations taking small steps forward for the promotion of peace. An opportunity from the establishment of transboundary conservation site is the resolution of conflicts and the promotion of peace. Borders are often associated with conflicts, but when the strict political notions of borders are blurred, allowing movement of people and wildlife, it also helps in alleviating the history of conflict. There are two different kinds of conflicts that can occur in conservation zones and borderland areas. The first type is the human–wildlife conflict and the other type is human–human conflicts. Due to the restriction of animal migration across a fenced border, the pressure on and from the people living in the vicinity of forest areas create human–wildlife conflicts. Conflicts also arise when both human and wildlife rely on the same resource, for example, a pond or share the same area for living and migrating. These conflicts can be resolved by either giving an alternative space to either humans or animals. The opening of borders for either of the stakeholders make the possibility of resolving the conflicts easier (Murphy, 2008). In the case of the KAZA transboundary region discussed above, a part of the region is developed as a chilly growing market which deters elephants who would often conflict with about 2000 people who live on their migratory corridor. Lastly, transboundary conservation through the recognition of joint work across the border can help in promoting peaceful relations between two or more countries. It can also help to commemorate long-standing peaceful relationships, and the historical significance of the region where a battle might have been fought, or a border issue might have been resolved. It can also be established initially for conservation purposes but in future it has the potential to urge the nations to resolve border conflicts through environmental management (Braak et al., 2006).

3.5 Under What Circumstances Can Transboundary Conservation Achieve These Opportunities?

A successful transboundary conservation approach has the potential to provide multifaceted ecological, social, economic, political and cultural benefits, but only if it is planned and implemented successfully. If it is flawed in specific ways, then it can worsen the status quo. These inefficiencies can result in further biodiversity loss, breach of human rights, violence and political turmoil. Several researchers have investigated pre-existing or potential transboundary conservation areas to understand the circumstances needed to achieve the associated opportunities. The following paragraphs discuss the different factors that can impact the implementation of a transboundary conservation approach and how these factors should be considered to minimize the negative impact and maximize the realization of these benefits. These observations and results come from all around the globe from varied types of parks having different purposes for park formation. There can be no one way of implementing transboundary conservation approach as different places differ not only in their ecology, but also in their social, cultural and political makeup which in turn affects the needs for their establishment. Let us have a look at some of the circumstances for a holistically designed and successful transboundary conservation framework.

3.5.1 Communication Is Key

For any successful project where several people are working together, the different partners/teams/collaborators must interact regularly. It requires that the flow of information be transparent, timely and well-coordinated. These requirements are also valid for a successful conservation plan. And, more so for a successful transboundary conservation approach as two or more countries are involved in it. Often, the biggest hindrance in the formation of transboundary conservation could be that the border nations are not in communicative terms. And their communication becomes even more critical after the formation of a transboundary conservation site. A transboundary conservation site has two or more stakeholders as compared to a single nation/state protected area. The way they communicate can drastically affect the level of implementation of the transboundary conservation approach.

Transboundary conservation planning requires all the participatory nations to collaborate, communicate and engage at a regular basis during all the steps of planning. A study (McCallum et al., 2015) conducted on all the transboundary conservation areas in Central America, South America and the Caribbean showed that among the various benefit the highest benefits were reported to the improvements to biodiversity which was directly correlated with improved communication. Improved communication between the partners was associated with improved spatial, management and socioeconomic benefits. For all the transboundary parks studied, the catalyst for the formation was spatial suitability. Still, the collaboration increased communication frequency which showed a positive correlation with improved biodiversity. Increased cooperation led to increased revenue from the sources of tourism and government funding. There was also a significant increase in NGO funding, followed by the generation of revenue through cultural events and exchanges. The increased revenue has the potential to supplement further conservation work. A more meaningful and enhanced communication also showed a direct correlation with joint biodiversity management, bio-threat mitigation and socioeconomic activities. It is possible because improved communication facilitates improved response to and management of threats over a larger area with combined resources of the transboundary partners. In this way, effective communication promotes the potential benefits from transboundary conservation approach; however, communication alone is not the only aspect that can bring these benefits. When communication is firm and smooth, the pitfalls can be identified, and with the willingness of partners, these can be overcome. Thus, communication might not be sufficient all by itself, but when communication is secure, it can help to strengthen all other factors for transboundary conservation.

3.5.2 The Cost of Conservation

In the most basic economic senses, a project is feasible when its benefits outweigh the costs. So, naturally, a transboundary conservation implementation is also dependent on how much it costs as compared to what benefits it gives. The costs of transboundary conservation are of two types – the impacts of conservation on the socioeconomic aspects and the cost of enforcement (Ban & Klein, 2009). Conservationists aim to minimize the cost of establishing a transboundary conservation area while achieving maximum possible benefits that are primarily associated with representing the targets for the designated area (Stewart & Possingham, 2005). It is essential to recognize and incorporate the costs in conservation planning majorly for two reasons. First, because it helps in reducing conflicts with socioeconomic actors by minimizing the impacts resource users would face due to conservation planning (Klein et al., 2008a, 2008b). And secondly, it helps in increasing the feasibility and manageability of the plan by making it cost-effective (Carwardine et al., 2008; Naidoo et al., 2006). Let us take an example from a prospective co-managed area between the USA and Mexico in the Baja California region (Arafeh-Dalmau et al., 2017). The researchers identified that if a network or transboundary area is established, there would be high costs to the fishermen due to increased fishing efforts and reduced commercial landings. The fishers might have to travel greater distances which will increase their effort for which the commercial scale of business should be met. This socioeconomic cost could be balanced by the possible spillover effect (Halpern et al., 2009) of improved fishing resources outside of the reserve due to improved conservation of priority breeding grounds, for instance. On the other hand, the enforcement costs could emerge from building the surveillance capacity, the type of coast, distance travel based on the size of the reserve, availability of resources for surveillance for all the parties and so on. In this case, the enforcement cost would be lowest for the party that already has the surveillance resources and are operating in an area with more boats per area, which has to travel a shorter distance and has higher cliffs (that will provide a better view for surveillance). In this way, the aim of the stakeholders should be to design such a plan that has low socioeconomic and enforcement costs while giving maximum benefits to increase its feasibility.

3.5.3 Governance

A transboundary conservation approach involves multiple levels of political organization for its implementation, and it also creates another layer of governmental involvement at the international level. The way different layers of government are involved, who is involved, how well they engage, manage and understand the needs of the project becomes key to the success of a transboundary conservation approach. Earlier in this chapter, different levels and types of governance approaches to transboundary conservation were introduced. These are valid options, but the best one depends on the objectives of the plan and the needs of the region. Often a hybrid approach to governance (Miller et al., 2019) gives the best solution. Some of the flaws due to ill-planned governance and possible opportunities for improvements are discussed in this section.

The location of decision-making and the scale of involvement are two critical components of the governance of transboundary conservation. Faure et al. (2010) show that for a transboundary conservation approach to generate potential benefits, the governance should be designed at a level to meet the highest benefits to the public. The different levels of governance that are needed to meet the defined goals of a specific project should be identified and included in the plan. It should also be made sure that different levels of political organization from one country matches or complements that of the other country. For example, in India, the conservation of a forest is not only dependent on the national level policies but also on the individual state's jurisdiction. If India were to sign a transboundary management plan with Bangladesh, there the national level policy is the only policy that guides forest management, and the roles of individual forest divisions might only be to the level of implementation. In this case, if the transboundary conservation panel involved national level stakeholders from Bangladesh, then similar representation from both the national level and state level from India needs to be involved in balancing the authority.

However, there are implications of involving the state power, especially when they are given the utmost authority as in the case of a top-down governance approach. Duffy (2006) showed that in a top-down governance approach, the top-most layer of the governance is given so much authority over the management that the effectiveness of the plans is compromised as the top-down approach to transboundary conservation tends to promote violence, dislocations and atrocities. The comparison of the transboundary approach with colonial conservation strategies and drawing similarities between them highlights the breaching of human rights that a top-down transboundary approach could cause (Duffy, 2006).

Such turmoil, violence and injustice were seen after the formation of the Limpopo National Park in Mozambique. When the fencing border between the Limpopo National Park and Kruger National Park in South Africa was removed to aid elephant migration between the two parks, the decision lacked community consultation with about 6,000 locals who lived on the Mozambique side of the territory. The community only got to know about it when elephants started migrating, and there was human–wildlife conflict as the community was enraged as the planners did not consider the threats to their livelihoods and in return threatened to kill the wildlife (IUCN-ROSA, 2002: 6–7).

While still criticizing the poor governance structure of transboundary conservation plans, King and Wilcox (2008) blame the inefficiencies on the excessive involvement of international organizations and NGOs in promoting, funding and planning these schemes without paying attention to the local context. They argue that transboundary conservation leads to 'minimise (d) political context', 'hegemony of international conservation agendas' while promoting 'economic neoliberalism'. They recognize the benefits to biodiversity but criticize that these plans are socially and politically ignorant, which underscores or worsens the status quo.

Additionally, the changes in global politics and the changes in politics within the nation-states bring in another added layer of governance-related uncertainty. This uncertainty can impact the long-term benefits of the transboundary conservation approach when the plan is not resilient and adaptive. In this way, the transboundary conservation approach is not immune to such global and local changes. The way these changes are absolved and acted upon by one nation is not like that of other participating nations which add on to the pressure. Other governance-related factors are related to the illegal activities that are commonly practised in borderland areas. These can spring from the trade networks on the border, or poaching, mainly when one nation's policy to deal with these activities is different from the others, it brings in possible avenues for conflict.

Kark et al. (2015) and Linde et al. (2002) propose that representation of both government and locals should be there at all the levels and risk assessment should be done at multiple levels while involving stakeholders from across the scale. This would ensure that the challenges of different scales involved in a transboundary conservation approach are minimized. Here, the understanding of power relations between each level, within and between participating countries, is also essential. Lessons from a successful and failed inclusion of the local communities and their implications on the effectiveness of the transboundary approach can be seen from Central and South East Asia. In the Pamir and Pamir-Alai mountains of Kyrgyzstan and Tajikistan, organic levels of community involvement from both the countries in resource management were present even before the implementation of a structured transboundary approach. Even after the implementation of a structured transboundary approach, a large part of decision-making was done at the local level where the community was involved and empowered. Whereas, in the Heart of Borneo project involving collaboration between Malaysia, Indonesia and Brunei, the involvement of the community was of restrictive and merely at the representative level.

In line with the same, Schoon (2013) emphasizes that the government should not be either top-down or bottom-up. But, the opportunities for combing both approaches into a hybrid approach might be a better solution. However, there cannot be one elixir solution to the global transboundary conservation planning, but the plan needs to start with a thorough understanding of multiple stakeholders. In the Kgalagadi Transfrontier Park in the Kalahari deserts of Botswana and South Africa, a bottom-up institutional approach to transboundary management was taken where the ground officials adaptively worked together to address the day-to-day challenges of management. This is in contrast to the selective top-down approach in The Great Limpopo Park discussed above, which is a park imposed upon the local community and also the local officials. These different starting points to planning for the transboundary approach resulted in differing capacities to improvement in the future and made their management path-dependent. Schoon (2013) shows how the point of the start or the 'location of decision-making' as discussed above plays an influential role in deciding the future of the shared territory. While a top-down is needed to bring together the nations at an international level to enact a treaty for park formation and establish communication and policies, operational development at the bottom-up level is needed in the future to strengthen the implementation and functioning of the territory. Thus, with a combination of both a top-down and bottom-up approach hybridity in co-governance can be achieved, which transboundary commons researchers have proclaimed as a practical solution (Miller et al., 2019).

3.5.4 Regions as Social Practice – Other Factors Impacting Effective Transboundary Conservation

In addition to improved governance, enhanced communication and reduced costs, there are several other factors on which the effectiveness of transboundary conservation depends. Paasi (2002: 200) reflects on the notion of border areas in Europe as 'regions on paper' which are far away from regions as a social practice. These other factors arise from transboundary regions not being part of the social arena but only being called as the unified region in management plans. A research conducted by Trillo-Santamaría and Paül (2016) shows how most of their interviewees who were stakeholders of transboundary management referred to the region only concerning the part from their country, but not to the extent of the unified area. Even among those who directly engage with how an area would be managed do not truly consider that area as a unified region but as one made up of different parts of the one to which they belong. In other words, they did not consider themselves as 'all' but 'one of all'.

Due to this 'region in discourse but not in practice' notion, there are several other factors that emerge. One of these impacts is reduced involvement and participation of the local community as stakeholders. They are merely involved in a few steps of public consultation, but often they might just be briefed about the oncoming projects without involving them or their knowledge. As multiple local communities are present as prospective stakeholders to transboundary conservation, their voice, feelings and knowledge should be involved as a central theme. Next, the planning process needs to be iterative and evolving by constant engagement of and communication between the stakeholders, but not as a sequential and straightforward process in which there is no scope to adapt. Third, land use planning should be thoroughly studied, understood and applied with care so that possible venues and subjects of conflicts from improper land use and resource planning can be avoided. Fourth, stakeholders should identify their role not merely as planners or managers but as those who are parts of the functioning of the TBPA where constant monitoring, observation and planning need to be applied to the functioning of the transboundary region. Fifth, although funding is essential and central to the implementation of the framework, it is not the most crucial element (Trillo-Santamaría & Paül, 2016). Without a joint strategic plan, and coordinated functioning, the funding obtained for the project can just be accessed and wasted on individual country projects which are not answering large-scale objectives of the project. Therefore, planners should not concentrate their attention on the attainment of the fund. But they should be concentrated towards a through plan which gives importance to historic administrative, legal, planning, cultural and political systems of the bounded nations as these characteristics do not disappear even after integrating the system.

3.5.5 Finding Common Ground

Given the challenges of a transboundary conservation approach, a common ground needs to be established to overcome these challenges so that the opportunities of the approach can be enjoyed. However, there cannot be one defined solution to the problem due to the uniqueness of geography, politics, society and culture around the world. Mason et al. (2020), while studying the feasibility of transboundary conservation at a global scale used three parameters to select areas that are more suitable to give higher benefits through transboundary conservation. They defined feasibility on the bases of the level of engagement (Boschee et al., 2015) governance indicators (Kaufmann et al., 2011) and human pressure (Venter et al., 2016). If a border has a higher level of collaboration and engagement, a better indicator of good governance and low human pressure than that area was given higher feasibility over other areas where one or more of the indicators scored low. Here, works from other researchers as mentioned above is in agreement with using a higher level of engagement and better governance has positive indicators of higher feasibility; however, Mason et al.'s (2020) claim that areas with low human pressure are more feasible for the transboundary approach is contested. There are two claims in favour and against the use of low human pressure as an indicator of a more feasible area for international conservation collaboration. It is argued that areas with low human pressures would have a healthier ecosystem that will not give a significant justification to put funds for conservation when those funds can be used at other places that are in a dire need for conservation. But on the other hand, it is easier to manage areas with lower human pressure due to lower socioeconomic and enforcement costs, which makes it easier to attain other benefits that a transboundary approach could generate.

Despite the disagreement with one of the parameters, the kind of quantitative evaluation of feasibility done in the study by Mason et al. (2020) is a good approach to consider when planning transboundary conservation. Planners should understand the different indicators of collaboration across the border not only to score the level of engagement but also to identify possible shortcomings that can be included in the plan for improvement. Further, this score can be used as a baseline to be compared while accessing the effectiveness of the implemented plan. These parameters should be analyzed before implementation and at multiple stages to attain the highest levels of a well collaborated transboundary governance.

In the above sections, it was highlighted that a transboundary approach comes with various opportunities that can be outweighed by the associated challenges if it is not planned well. And it was also shown that different areas could not be considered similar. Thus, a proposed way of planning transboundary conservation can be an adaptive one. In which you start by establishing the baselines of the territory in question, define short-term objectives that are attainable but also long-term objectives to define the path forward. And then, these objectives, the plan and the progress are revisited at periodic terms to learn from what worked well, what did not and design improvements based on these lessons. This kind of adaptive governance or adaptive management of transboundary areas helps in modifying the plan along the way and makes it easier to include different stakeholders at an organic level of planning. Ultimately, an adaptive approach helps with increased capacity building and a more robust response to problems. Thus, by acknowledging the complexity of transboundary areas, an adaptive approach would 'strengthen institutional arrangements in ways that build robustness and promote long enduring institutions' (Schoon, 2013).

References

Arafeh-Dalmau, N., Torres-Moye, G., Seingier, G., Montaño-Moctezuma, G., & Micheli, F. (2017). Marine spatial planning in a transboundary context: Linking Baja California with California's network of marine protected areas. *Frontiers in Marine Science*, *4*. https://doi.org/10.3389/fmars.2017.00150

Araújo, M. B., Cabeza, M., Thuiller, W., Hannah, L., & Williams, P. H. (2004). Would climate change drive species out of reserves? An assessment of existing reserve-selection methods. *Global Change Biology, 10*(9), 1618–1626. https://doi.org/10.1111/j.1365-2486.2004.00828.x

Ban, N. C., & Klein, C. J. (2009). Spatial socioeconomic data as a cost in systematic marine conservation planning. *Conservation Letters, 2*(5), 206–215. https://doi.org/10.1111/j.1755-263X.2009.00071.x

Boschee, E., Lautenschlager, J., O'Brien, S., Shellman, S., Starz, J., & Ward, M. (2015). ICEWS coded event data. *Harvard Dataverse, V28,* UNF:6:NOSHB7wyt0SQ8sMg7+w38w== [fileUNF]. https://doi.org/10.7910/DVN/28075

Braack, L., Sandwith, T., Peddle, D., & Petermann, T. (2006). Security considerations in the planning and management of transboundary conservation areas. IUCN. Retrieved February 11, 2015, from https://portals.iucn.org/library/efiles/documents/2006-056.pdf

Carroll, C., Dunk, J. R., & Moilanen, A. (2010). Optimising resiliency of reserve networks to climate change: Multispecies conservation planning in the Pacific Northwest, USA. *Global Change Biology, 16*(3), 891–904. https://doi.org/10.1111/j.1365-2486.2009.01965.x

Carwardine, J., Wilson, K. A., Watts, M., Etter, A., Klein, C. J., & Possingham, H. P. (2008). Avoiding costly conservation mistakes: The importance of defining actions and costs in spatial priority setting. *PLOS ONE, 3*(7), e2586. https://doi.org/10.1371/journal.pone.0002586

Casagrandi, R., & Gatto, M. (1999). A mesoscale approach to extinction risk in fragmented habitats. *Nature, 400*(6744), 560–562. https://doi.org/10.1038/23020

Chan, K. M. A., Satterfield, T., & Goldstein, J. (2012). Rethinking ecosystem services to better address and navigate cultural values. *Ecological Economics, 74*, 8–18. https://doi.org/10.1016/j.ecolecon.2011.11.011

Christie, P. (2004). Marine protected areas as biological successes and social failures in Southeast Asia. *American Fisheries Society Symposium, 42*, 155–164.

Dudley, N., Groves, C., Redford, K. H., & Stolton, S. (2014). Where now for protected areas? Setting the stage for the 2014 world parks congress. *Oryx, 48*(4), 496–503. https://doi.org/10.1017/S0030605314000519

Duffy, R. (2006). The potential and pitfalls of global environmental governance: The politics of transfrontier conservation areas in Southern Africa. *Political Geography, 25*(1), 89–112. https://doi.org/10.1016/j.polgeo.2005.08.001

Faure, M., Goodwin, M., & Weber, F. (2010). Bucking the Kuznets curve: Designing effective environmental regulation in developing countries. *Virginia Journal of International Law, 51*, 95.

Ferrier, S. (2002). Mapping spatial pattern in biodiversity for regional conservation planning: Where to from here? *Systematic Biology, 51*(2), 331–363. https://doi.org/10.1080/10635150252899806

Graham, J., Amos, B., & Plumptre, T. (2003). *Governance principles for protected areas in the 21st century, 50.*

Halpern, B. S., Lester, S. E., & Kellner, J. B. (2009). Spillover from marine reserves and the replenishment of fished stocks. *Environmental Conservation, 36*(4), 268–276.

IUCN-ROSA. (2002). *Rethinking the great Limpopo transfrontier conservation area and TBNRM developments in Southern Africa: A discussion paper for a collaborative workshop to establish current baseline data and current research efforts for TBNRM Management in Southern Africa.* IUCN-ROSA, Harare: Southern Africa Wildlife College.

Jennerjahn, T. C., Gilman, E. L., Krauss, K. W., Lacerda, L. D., Nordhaus, I., & Wolanski, E. (2017). *Mangrove ecosystems under climate change.* https://link-springer-com.libproxy1.nus.edu.sg/chapter/10.1007%2F978-3-319-62206-4_7

Kark, S., Tulloch, A., Gordon, A., Mazor, T., Bunnefeld, N., & Levin, N. (2015). Cross-boundary collaboration: Key to the conservation puzzle. *Current Opinion in Environmental Sustainability, 12,* 12–24. https://doi.org/10.1016/j.cosust.2014.08.005

Kaufmann, D., Kraay, A., & Mastruzzi, M. (2011). The worldwide governance indicators: Methodology and analytical issues. *Hague Journal on the Rule of Law, 3*(2), 220–246. https://doi.org/10.1017/S1876404511200046

King, B., & Wilcox, S. (2008). Peace parks and jaguar trails: Transboundary conservation in a globalising world. *GeoJournal, 71*(4), 221–231. https://doi.org/10.1007/s10708-008-9158-4

Klein, C. J., Steinback, C., Scholz, A. J., & Possingham, H. (2008a). Effectiveness of marine reserve networks in representing biodiversity and minimizing impact to fishermen: A comparison of two approaches used in California. *Conservation Letters, 1*(1), 44–51. https://doi.org/10.1111/j.1755-263X.2008.00005.x

Klein, C., Chan, A., Kircher, L., Cundiff, A. J., Gardner, N., Hrovat, Y., Scholz, A., Kendall, B. E., & Airamé, S. (2008b). Striking a balance between biodiversity conservation and socioeconomic viability in the design of marine protected areas. *Conserv Biol, 22*(3), 691–700. https://doi.org/10.1111/j.1523-1739.2008.00896.x

Mason, N., Ward, M., Watson, J. E. M., Venter, O., & Runting, R. K. (2020). Global opportunities and challenges for transboundary conservation. *Nature Ecology and Evolution, 4*(5), 694–701. https://doi.org/10.1038/s41559-020-1160-3

Mbora, D. N. M., & McPeek, M. A. (2010). Endangered species in small habitat patches can possess high genetic diversity: The case of the Tana River red colobus and mangabey. *Conservation Genetics, 11*(5), 1725–1735. https://doi.org/10.1007/s10592-010-0065-0

McCallum, J. W., Vasilijević, M., & Cuthill, I. (2015). Assessing the benefits of transboundary protected areas: A questionnaire survey in the Americas and the Caribbean. *Journal of Environmental Management, 149,* 245–252. https://doi.org/10.1016/j.jenvman.2014.10.013

Miller, M., Middleton, C., Rigg, J., & Taylor, D. (2019, July). Hybrid governance of transboundary commons: Insights from Southeast Asia. *ResearchGate.* https://www.researchgate.net/publication/334346247_Hybrid_Governance_of_Transboundary_Commons_Insights_from_Southeast_Asia

Murphy, C. A. (2008). Living in a large-scale commons—The case of residents of a national park in the Kavango-Zambezi transfrontier conservation area (KaZa TFCA), Southern Africa. Paper delivered at the IASC Conference, 14–18 July 2008, Cheltenham, England.

Naidoo, R., Balmford, A., Ferraro, P. J., Polasky, S., Ricketts, T. H., & Rouget, M. (2006). Integrating economic costs into conservation planning. *Trends in*

Ecology and Evolution, *21*(12), 681–687. https://doi.org/10.1016/j.tree.2006.10.003

Salisbury, D. S., López, J. B., & Vela Alvarado, J. W. (2011). Transboundary political ecology in Amazonia: History, culture, and conflicts of the borderland Asháninka. *Journal of Cultural Geography*, *28*(1), 147–177. https://doi.org/10.1080/08873631.2011.548491

Sandwith, T., Shine, C., Hamilton, L., Sheppard, D., & Phillips, A. (2001). *Transboundary protected areas for peace and co-operation.* IUCN, Gland, Switzerland, and Cambridge, UK.

Sayer, J., Sunderland, T., Ghazoul, J., Pfund, J.-L., Sheil, D., Meijaard, E., Venter, M., Boedhihartono, A. K., Day, M., Garcia, C., van Oosten, C., & Buck, L. E. (2013). Ten principles for a landscape approach to reconciling agriculture, conservation, and other competing land uses. *Proceedings of the National Academy of Sciences*, *110*(21), 8349–8356. https://doi.org/10.1073/pnas.1210595110

Schoon, M. (2013). Governance in transboundary conservation: How institutional structure and path dependence matter. *Conservation and Society*, *11*(4), 420. https://doi.org/10.4103/0972-4923.125758

Stewart, R. R., & Possingham, H. P. (2005). Efficiency, costs and trade-offs in marine reserve system design. *Environmental Modeling and Assessment*, *10*(3), 203–213. https://doi.org/10.1007/s10666-005-9001-y

Trillo-Santamaría, J.-M., & Paül, V. (2016). Transboundary protected areas as ideal tools? Analysing the Gerês-Xurés transboundary biosphere reserve. *Land Use Policy*, *52*, 454–463. https://doi.org/10.1016/j.landusepol.2015.12.019

van der Linde, H., Oglethorpe, J., Sandwith, T., Snelson, D., Tessema, Y., Tiéga, A., & Price, T. (2002). *Beyond boundaries: Transboundary natural resource management in sub-Saharan Africa.* https://www.cabdirect.org/cabdirect/abstract/20023197241

Vasilijević, M., Zunckel, K., McKinney, M., Erg, B., Schoon, M., Rosen Michel, T., Groves, C., & Phillips, A. (2015). *Transboundary conservation: A systematic and integrated approach.* International Union for Conservation of Nature. https://doi.org/10.2305/IUCN.CH.2015.PAG.23.en

Venter, O., Sanderson, E. W., Magrach, A., Allan, J. R., Beher, J., Jones, K. R., Possingham, H. P., Laurance, W. F., Wood, P., Fekete, B. M., Levy, M. A., & Watson, J. E. M. (2016). Global terrestrial human footprint maps for 1993 and 2009. *Scientific Data*, *3*(1), 160067. https://doi.org/10.1038/sdata.2016.67

West, P., Igoe, J., & Brockington, D. (2006). Parks and peoples: The social impact of protected areas. *Annual Review of Anthropology.* https://www-annualreviews-org.libproxy1.nus.edu.sg/doi/abs/10.1146/annurev.anthro.35.081705.123308

Westing, A. H. (1998). Establishment and management of transfrontier reserves for conflict prevention and confidence building. *Environmental Conservation*, *25*(2), 91–94. https://doi.org/10.1017/S0376892998000137

4 Riverbank Erosion and Inter-Community Relationships in Majuli

Political Implications of a Changing Landscape in Assam

Avijit Sahay

4.1 Introduction

A natural hazard is defined as any sudden environmental event that causes or has the capacity to cause damage to life or property or both. Thus, hazards by their very nature affect the material conditions of life, and as such operate at a level of political discourse (Pelling, 1999). More importantly, the impact of hazards has been shown to be proportional to vulnerability of an individual, group or society, which itself is intrinsically linked to the economic (Birkmann & Wisner, 2006) and political (Kasperson & Kasperson, 2005; Oliver-Smith & Hoffman, 1999) marginalization of particular groups, and thus, depends upon the access to rights, resources and assets (Blaikie et al., 2014; Sen, 1981). The concepts of marginalization and vulnerability are therefore linked to each other in hazard studies, as the least powerful groups are most vulnerable to hazards because of their limited economic and geographical options (Collins, 2008) and are thus further marginalized to live and work in degraded landscapes and occupying hazardous environments (Robbins, 2004).

It is generally believed that in the face of a common threat from natural calamities, societies come together and conflicts are weakened. Ian Kelman (2003) considers disasters to be facilitators of diplomatic efforts to end conflicts in any region. Similarly, some more works have shown disasters to lead to the growth of altruistic or therapeutic communities (Quarantelli and Dynes, 1976) that have led to a dampening of conflicts. However, from the above discussion on the nexus between disasters, vulnerability and marginalization, it is clear that natural hazards can disrupt social equilibrium in communities by intensifying the latent vulnerabilities of the social groups. Oliver-Smith (1990) has shown how human societies are all stratified and this becomes even more pronounced during and after a disaster. Other

DOI: 10.4324/9780367486433-4

works have also identified how disasters intensify pre-existing status differences and inequalities (Haas, Kates & Bowden, 1977; Peacock & Bates, 1982; Bolin & Bolton, 1986) and thereby give rise to conflicts. Similarly, Sorokin (1946) and Cuny (1994) have shown how disasters can cause latent divisions in societies to become major conflicts.

Thus, natural hazards can have profound political, social and economic impacts. It is thus clear that disasters can have both a dampening as well as an intensifying effect on social relations in disaster zones. The exact effect that disaster produces therefore depends on many different variables like pre-existing grievances, structural inequalities, resource scarcities, vulnerabilities and the scale of marginalization that communities face. The current study attempts to understand the issues related to such factors in Majuli island of Assam and the impact this has on social relations.

4.2 Riverbank Erosion in Majuli

Majuli is an island on River Brahmaputra in the Indian state of Assam, covering approximately 584 km² and home to 167,304 persons. It is formed when an anabranch of Brahmaputra, locally called Kherkutia Xuti or River Luhit, separates from the main river, moves north, travels almost 100 km as a separate river and finally merges back with the main river near Golaghat district. The district of Majuli lies between the districts Lakhimpur on north bank of Brahmaputra and Jorhat on the south bank and is shown in Figure 4.1.

The Brahmaputra Valley in Assam has witnessed frequent earthquakes, with the earthquakes of 1897 and 1950 being particularly severe. The 1950 earthquake measured 8.7 on the Richter Scale and disturbed the geology of the river channel, affecting the gradient of this river, stopping the flow temporarily and bringing about flooding and rapid accumulation of enormous volumes of sediment in the channel.

Post 1950, the River Brahmaputra in the south and Rivers Luhit and Subansiri in the north have been continuously changing their course and thereby eroding land in Majuli, leading to almost one-third to two-thirds of the area of the island being lost to the river. The loss of land because of erosional work of the river can be understood by the help of remote sensing images of Majuli from 1975 (Figure 4.2), 2000 (Figure 4.3) and 2015 (Figure 4.4).

This change in the morphology of the Brahmaputra has resulted in loss of land, which can be expressed in terms of the number of villages that have been wiped out by the river in the last six decades. There are a total of 243 villages in Majuli, out of which 107 have been totally or partially been lost to the river. The island of Majuli is subdivided into three *Mauzas*

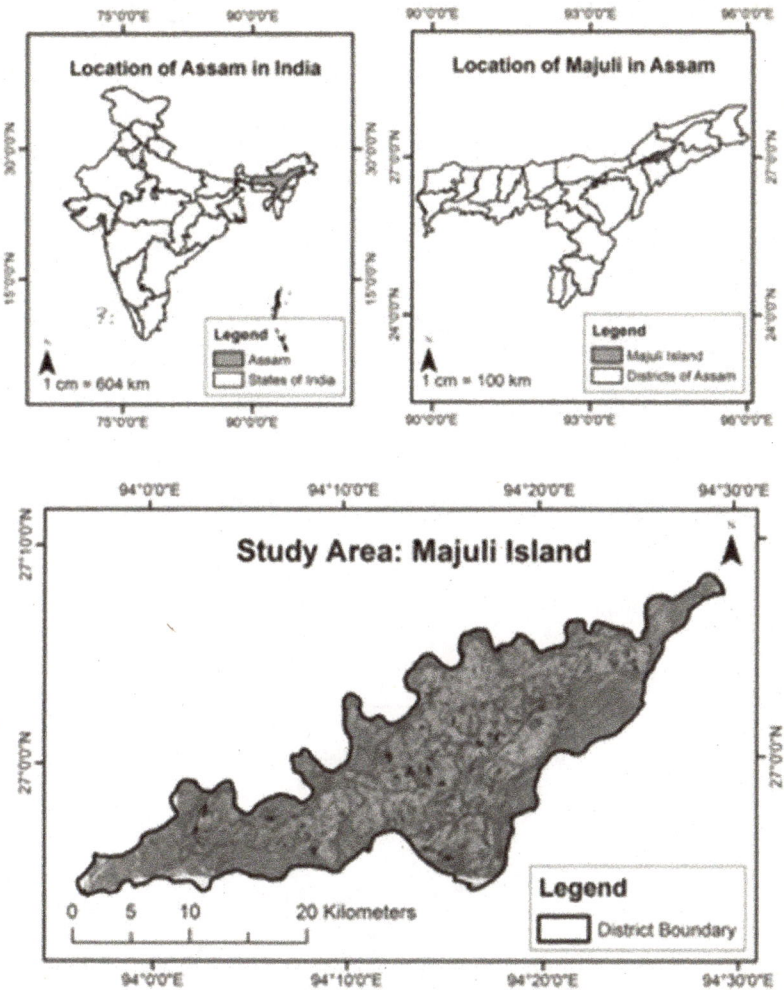

Figure 4.1 Location of Majuli

(revenue collection units): Ahataguri in the south east, Kamlabari in the centre and Salmora in the north west as given in Figure 4.5. From the remote sensing images, it can be clearly seen that Ahataguri has been completely lost to the river, and currently, maximum erosion is taking place in Salmora. Kamlabari, though suffering erosion, has not been affected to the same extent as Ahataguri and Salmora. Thus, erosion happens in specific

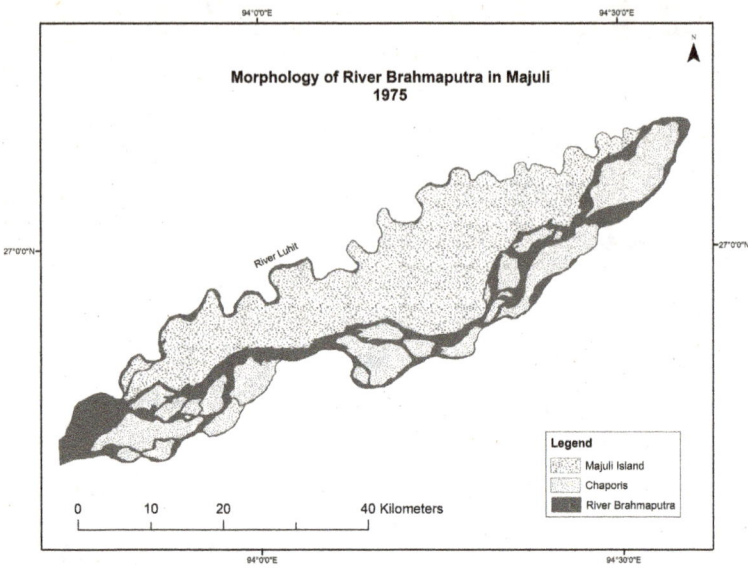

Figure 4.2 Majuli in 1975

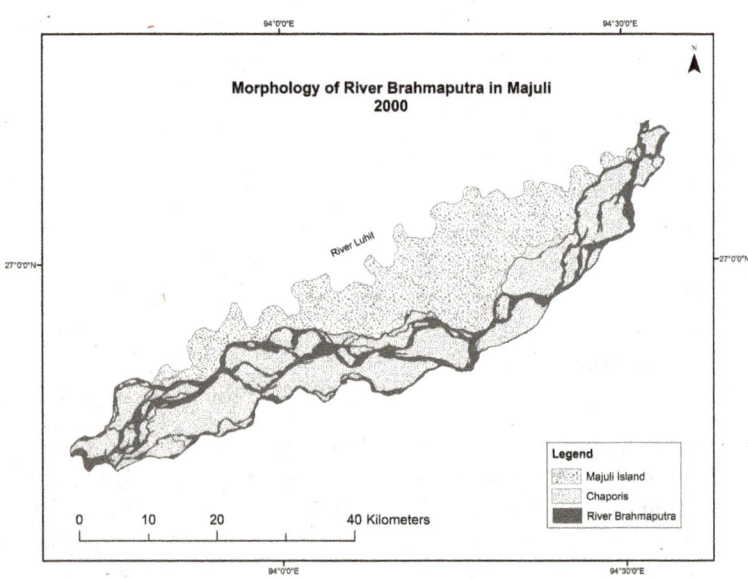

Figure 4.3 Majuli in 2000

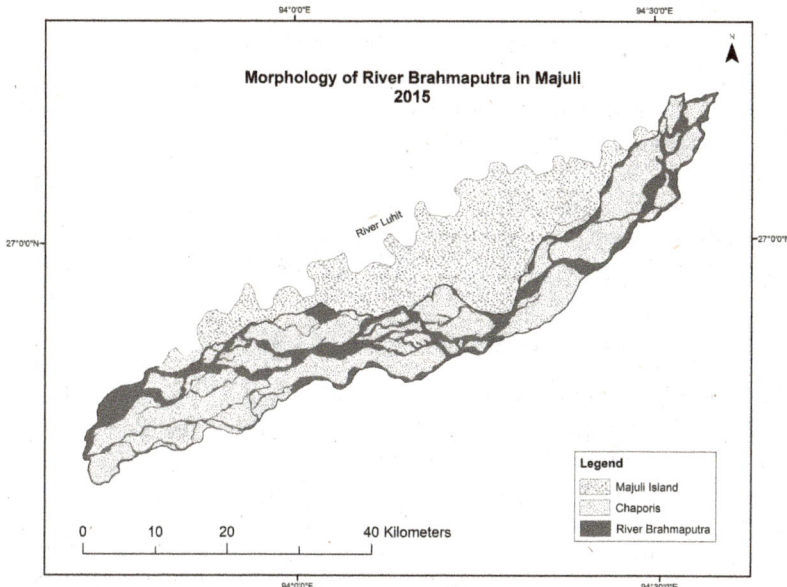

Figure 4.4 Majuli in 2015

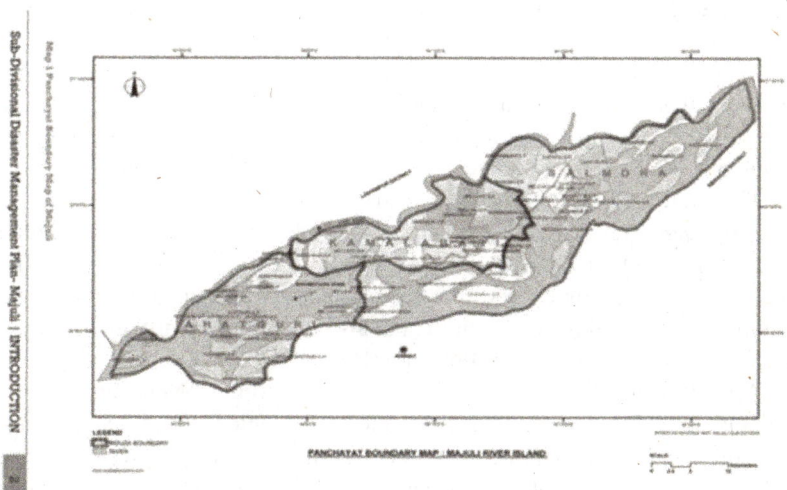

Figure 4.5 Majuli: Administrative Divisions

geographic locations, which makes people or social groups living here more vulnerable than the rest.

The most significant impact of riverbank erosion has been on the livelihood pattern of the island. For centuries, the occupation structure of the island has been dominated by the twin resources of fishing and agriculture. However, the changing course of River Brahmaputra, as well as the associated erosion of agricultural land, has affected both these resources. Moreover, as discussed above, the erosion happens in special geographical locations in the island which means that the impact of erosion is felt unequally both geographically as well as demographically.

4.3 Social and Cultural Landscape in Majuli

Majuli is a multiethnic society, chiefly comprising Assamese, Misings, Kacharris, Deoris and Nepalese, who have migrated to make Majuli their home. In spite of multiplicities of ethnicities, the religious beliefs of the island are shaped by the island's unique Vaishnavite culture called neo-Vaishnavism in Assam.

Neo-Vaishnavism was introduced in Majuli by the famous fifteenth-century social and religious reformer of Assam, Srimanta Shankardeb. Shankardeb, along with his disciple Madhavdeb promoted devotion to Lord Vishnu and his symbol Garuda, calling their new philosophy *Ek Sarna Hari Naam Dharma*, literally meaning complete surrender to One God (Choudhury, 2011). They banned idol worship and introduced many monastic elements in their new religious movement. Because of its opposition to idol worship, the Neo-Vaishnavite place of worship is devoid of all images and symbols and is simply called *naamghar*. The word *naamghar* is derived from the words *naam*, meaning name, and *ghar*, meaning house (Sahay & Roy, 2017). All prayers inside the *naamghar* are in the form of repeating the name of Lord Vishnu (Chaliha, 1998).

The monastic elements of Neo-Vaishnavism are found in the institution of *Satra*. The *Satras* are monasteries headed by the chief priest called *Satradhikar* and act as the core of all social and cultural life in Majuli. The *Satra* is deeply involved with the society, not just in regulating the social and religious life of the community, but also in preserving as well as propagating the neo-Vaishnavite philosophy (Baruah, 1994).

After the establishment of the first *Satra* in Majuli, Neo-Vaishnavism quickly spread to the rest of Assam and was adopted as the major socio-cultural and religious philosophy of the state. However, over the past 500 years, new philosophical currents were introduced in Assamese society, which meant that the hold of Neo-Vaishnavism began to loosen in much of Assam. But Majuli being an island and isolated from the rest of Assam,

Neo-Vaishnavism is preserved here in almost its pristine form. This extraordinary longevity was recognized by the state government when the Majuli Cultural Heritage Act was passed by the State Assembly of Assam in 2006, which granted Majuli the status of a Cultural Heritage Site of India (Sahay & Roy, 2017).

Majuli historically had 65 *Satras*, distributed across the island (Nath, 2009). However, because of riverbank erosion, many *Satras* too lost their land, which meant that while the larger and richer *Satras* could relocate, the smaller and poorer *Satras* closed down or simply moved out of Majuli. During my fieldwork in Majuli, I was informed that, currently, 22 *Satras* still remain active in Majuli. Out of these, some 8–10 *Satras* are big, while the rest are smaller, with some being only adhered to by a handful of families. For my fieldwork, I visited 18 of the remaining *Satras*. My major aim for this was to create a map of existing *Satras* in Majuli, and also to find out what role *Satras* played in helping those who had lost their land to the river and were affiliated with the *Satra*. The map that I created is shown in Figure 4.6.

Different tribal and ethnic groups that have migrated to Majuli were assimilated in the fold of *Satras*. The most important of these are the Mising people who now constitute more than 40% of the total population of Majuli. Originally belonging to the *Tani* group of tribes of Arunachal Pradesh, the Misings migrated to the plains of Assam and settled in districts of Upper Assam. Traditionally, the Misings practised nature and ancestral worship (Zaman, 2015) in the form of *Do Nyi: Po Lo* (The Sun and the Moon), but in Majuli, they quickly became members of different *Satras* and in the process adopted many Hindu beliefs and practices (Nath, 2009).

However, the process of assimilation has not been smooth, and in recent years there have been increasing clashes and conflicts between the Mising community and the Assamese community that is represented by the *Satras*. The Mising allege that in spite of accepting the Hindu faith, they are not given equal status in the communal and social life of *Satras* and are always treated as second-class citizens. The *Satras* on the other hand blame the Misings for not fully incorporating Hindu rites and rituals in their lifestyle. Two of the most important points of contention are Mising death rites and Mising food habits.

The Misings in Majuli still practise their traditional system of burying their dead as against cremating them according to *Satra* rituals. Also, the food and economy of Misings depend a lot on livestock rearing and therefore meat and especially pork constitute a big portion of the Mising diet. The *Satra* propagates vegetarian food habits and is critical of Mising people eating pork, which it anyway sees as dirty.

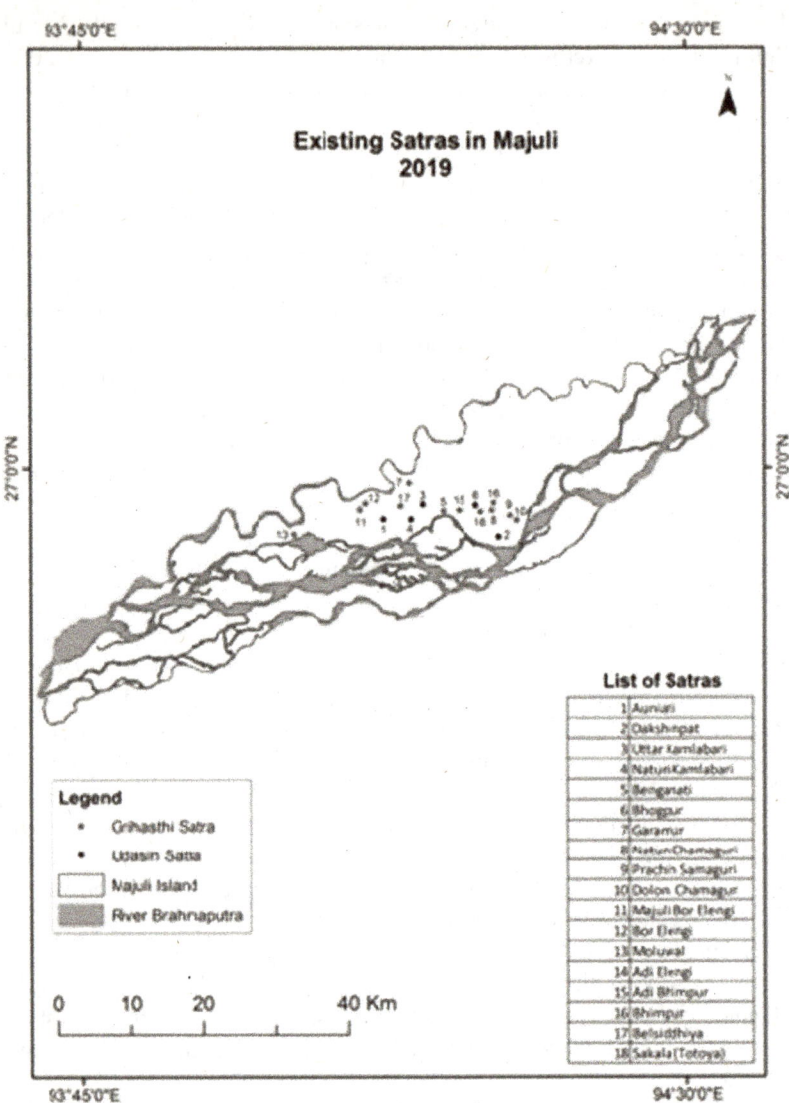

Figure 4.6 Majuli Satras

Thus, there already exists a fault line in Majuli society, in which Assamese community led by *Satras* and the tribal Mising community harbour grievance against each other. This problem is further exacerbated by the process of riverbank erosion and its unequal impact on the Assamese and Mising communities. The riverbank erosion in Majuli is discussed in subsequent paragraphs.

4.4 Impact of Riverbank Erosion on Different Communities

Mising and Assamese both constitute more than 40% each of the total population of Majuli. The Assamese community practises agriculture, while the Mising derive their income mostly from fishing, with some agriculture and animal rearing coexisting with fishing. Being a majorly fishing community, the Misings occupy the riverbank and are therefore more exposed to erosion and its effects than the Assamese. The Assamese community also live along riverbanks in some of the villages, but they mostly occupy the middle parts of the island, and are therefore not as vulnerable to erosion as the Mising.

The vulnerability of Mising people to riverbank erosion is well documented. Das (2014) has shown that wetlands along the riverbank are the most important sources of livelihood for Mising people in Majuli. These wetlands act as the primary source of fish for the community and also provide them with traditional food and medicinal plants like *Joloban, Laborua,* etc. *Gagola*, a type of reed used for making mats, is also found in these wetlands (Sarmah, 2012). The indigenous Mising mats are in great demand in Majuli, and now these are also being marketed by the Assamese government as handicrafts for a wider market in both Assam as well as the rest of India. Because of the importance of wetlands in Mising economic life, the community prefers to live along riverbanks. However, in times of riverbank erosion, this makes the Mising community particularly vulnerable to displacement.

While, it is clear that the Mising own less land and earn less income than the Assamese community in Majuli, it is also found that the average size of a Mising household is more than that of an Assamese household. While household size in Majuli is 5.03 persons per household, the Mising in Majuli have more than 7 persons per household (Das, 2013).

From the above discussion, it is clear that the Mising are more vulnerable to riverbank erosion both geographically as well as economically. However, even after being displaced by erosion, they are more marginalized in terms of getting relief. Das (2014) has shown that the Mising are least likely to be provided with land for resettlement and rehabilitation in Majuli. According to the Revenue Circle of Majuli, this is because Mising villages most vulnerable to riverbank erosion lie in wetlands as discussed earlier, and in Majuli, wetlands are classified as barren or waste land. When the government determines which lands should be compensated first, the preference is given to agricultural land, which is owned more by the Assamese community than the Mising (Das, 2014). Thus, Mising land is considered less important by the government. Because of this disparity in land allocation, a number of refugee settlement-like villages of the Mising tribe have sprung up along the main road in Kamlabari and Dakhin Kamlabari. The population living in these refugee settlements work as agricultural labourers in the land of those who remain unaffected by riverbank erosion on a 50:50 basis, wherein 50% of the produce of land is kept by the landowner and 50% is kept by the agricultural labourer (Sahay & Roy, 2017). Thus, because of riverbank erosion, the Mising community are being transformed from a landowning community to a landless labourer community.

To find out the impact of such vulnerability, displacement and marginalization of the Mising community on the inter-community relations in Majuli, fieldwork was conducted in Majuli. Riverbank erosion affected villagers from the Mising community and prominent Mising people were interviewed to find out their perception of *Satras* and the Assamese community. One of the methods employed was to attend *Do Nyi: Po Lo* ceremony, which is the traditional form of prayer in the Mising community and in which all the relatives, friends and high-ranking villagers are invited. Once the ceremony is over, some of the invitees are called upon to give a speech on community affairs. Sensing this as an opportunity, I wanted to attend *Do Nyi: Po Lo* in multiple villages during my stay. However, the ceremony is held to mark only the most auspicious events in Mising life and is therefore not held very often. Moreover, the participation in it is by invitation only, which is generally limited only to members of the Mising community. I had asked my host in Majuli to get me invited to as many events as possible, but unfortunately I could get an invitation for only one such event during my stay.

The *Do Nyi: Po Lo* that I attended was organized in Molual Miri village, falling under the Karatipar Gaon Panchayat. The village has more than 97% of the Mising population who claim allegiance to *Auniati Satra* but also practise all the traditional Mising rituals and beliefs. I was able to talk to two Mising community members during the ceremony and a summary of my interaction along with the name of the interviewee is given in the following paragraphs

4.4.1 Kamala Kaman, Retired from Indian Army, Garamur

We are originally a hill tribe from Arunachal Pradesh, but we came to Assam and now live in 11 plain districts of the state. One of our leaders, Parmanand, became a follower of Shankardeb's neo-Vaishnavism and we followed Parmanand into Vaishnav fold. Most of the Mising in Assam, and almost all of us in Majuli have taken Diksha in Satra and have become Hindus. But we do not follow most of the traditions of Hinduism. The Hindus have many different rituals, like they complete the death rituals in 13 days. However, because we bury our dead, we believe that it takes more than two years for the body to become a part of cosmos and so we have our death ritual between two to three years after death. The Satra has never been able to understand our ritual or accept it as some form of Hindu belief because they believe in a very rigid form of Hinduism. Again, all our religious ceremonies involve apang, a local alcoholic drink made from rice. We offer apang to our sun and moon gods along with pigs. The Satra look down upon our rituals because they consider one to be sin and the other to be dirty. Coming to erosion, you will find greatest impact of erosion has been felt by Mising people. We live on the north as well as the south banks and together around 40–50 Mising villages have been lost to the river. Comparatively, very few villages have been lost of the Assamese. However, when it comes to providing relief, government gives land first to Assamese, and then if any is left then it is provided to us. You can yourself see, if you drive from Kamlabari chariali towards Dakhinpat and Chamaguri Satra, you will see thousands of Mising houses on both sides of the road who just have a thatched house and a tin roof and are living as landless agricultural labourers. You will never find similar settlements of Assamese people because they have already been provided with land by the local authority. The Satra also never takes our side in any relief effort. They are mostly interested in preserving their own hegemony. There are some Christian missionaries who are now offering us help in exchange for accepting their faith and I don't blame those who have now converted because everyone is more interested in earning a living and having a house to live in. Overall, we feel that whatever adaptation was needed was done by Mising community but the Satras still are not accepting us as equals.

4.4.2 Nabo Kumar Gam, Principal, Rangachahi College, Majuli

Our problems with Satra are manifold. On one level, we feel that Satras want us to remain forever inferior to the Assamese community of Majuli. On the other hand, it is true that Misings are numerically more in Majuli now than the Assamese community and are therefore more vocal and assertive about our proper place in Majuli's society and culture. We are not against Satras, all we want is a more balanced relationship and an acceptance that even though our beliefs and rituals do not match totally with the Satras, we are still part of a composite culture of Majuli. Riverbank erosion has, however, further complicated the issue. Mostly, the government does not help us at all, but whatever help is forthcoming is given first to the Assamese community and then to us. This is in spite of the fact that Mising have been far more affected by erosion that the Assamese or any other community. While we feel aggrieved by this, we don't want conflict with any community. Ultimately, we are all a part of this island and the entire island is under threat because of riverbank erosion.

Apart from these, I also did a survey among the Mising population in Garamur and Kaniajan villages and the summaries of these interactions are given below.

4.4.3 Monjit Risong, Garamur

The root cause of all our problems with the Satras and, by extension the Assamese community, is that they want us to follow a very rigid religion and culture. We adopted Hinduism very late. Our chief gods are still the sun and the moon and we still pray to our ancestors. We do accept many rituals of Satras and take part in all religious ceremonies of Satras as well. But because of our adherence to our traditional way of life, we practise some things that Satras will never tolerate. For example, we eat pigs and perform a sacrifice of pigs in every ritual. The Satras believe in Vaishnavism and therefore they abhor meat and especially any kind of sacrifice or offering to gods. Also, a majority of Misings still bury their dead instead of cremating them. This is a big issue that we have with Satra. The Satras basically try to regulate the entire life of an individual from birth till death. The fact that we practise a ritual that is the complete opposite to what Satra wants us to practise is again a big problem. Many prominent Mising people have been criticizing the Satra and the way the Satras treat us as second class citizens. However, mostly the common Mising villagers do not want to confront the Satra. This is changing because of riverbank erosion. The Satra even though very powerful and rich, does not help people affected by riverbank erosion. In contrast to that, there are many Christian organizations working in Majuli now that promise to pay compensation and send our children to school in exchange of accepting Christianity. This situation is now being used by some of the Mising community leaders as a way to challenge the Satras and say to them that if you do not accept us as equals, then we will convert en masse to Christianity. Satras, on the other hand, instead of listening to Mising leaders are trying to expel Christian groups out of Majuli. Thus, there is an uneasy peace in the island between a group of Mising people on one hand and the Satras on the other hand.

4.4.4 Shyamanta, Kaniajan Village

I used to own 10 bighas of land in Bhokot Chapoori but the river eroded my entire village, I now live as a refugee in Kaniajan village and work in the fields of a person whose land lies in the middle part of Majuli and is therefore not affected by erosion. I get 50% of the harvest, while the rest is kept by the land owner. I have nine members in my family who depend on my meagre income and am therefore looking to go out of Majuli in search of better opportunities. When our land was eroded, we were not given any help by the government. The government has given some people land for resettlement in Majuli. Some more have been resettled out of Majuli in districts such as Jorhat, Golaghat and Lakhimpur. However, no one from my village got any government help. Even the Satras do not help us in any way. The Satras consider us unclean as we keep a lot of animals in our homes, including pigs. A friend of mine converted to Christianity because of such attitudes of Satras and now lives in Jengraimukh. His family was given money and their children attend English medium school in Lakhimi village. I also want to convert to Christianity but my family is against such a move.

4.4.5 Shankar, Kaniajan Village

I used to own land in Bhokot Chapoori but now my land and my home have been lost to the river. I live as a refugee in Kaniajan and work as an agricultural labourer in Garamur. Our family had owned land and lived in Bhokot Chapoori for hundreds of years, but in a matter of a few years, the river took all our land and made us landless. Every year we used to see the river coming closer to us with some of the villagers losing land and then shifting to Kaniajan annually. We used to meet and petition the government representatives but no one helped us and, ultimately, our entire land was lost to the river. The Satras are now very petty and do not concern themselves with our plight. They are interested in maintaining the purity of the Neo-Vaishnavite faith and according to them; we are the cause of impurity of their faith. It is because of this reason that I have stopped participating in Satra events. I have forbid my family from participating also. I know some of my friends have converted to Christianity but I do not intend to do so. I think our traditional religion was good for us and we should return to worshipping Do Nyi: Po Lo and our ancestors exclusively.

4.4.6 Amit Pegu, Sumoimari Village

I used to own a lot of land but all of it is now gone. Currently, I work in a tea stall that is owned by a man who lives in Garamur. Because his land was not eroded by the river, he employs a Mising man to work in his field and now runs this tea stall in Garamur. This is the reality of Majuli that some people now have double sources of income because the river did not take away their land, while some of us have been left with nothing. The hardest hit are Mising people because we are the people who have lost most of the land and consequently find ourselves to be refugees in our own land. I do not follow most of what our leaders say against Satras, but understand that they do not accept us as equals. Some of the people in our community say that we should now pay more attention to our traditional system of religion while others are saying we must remain part of Vaishnavism and Satras. I cannot care less what we are part of as long as someone helps us. This is true that Misings are given help only after Assamese people have been helped and this pains us. But this has always happened in Majuli and we are only now voicing our opposition to it.

4.5 Results and Discussions

From the above discussion, it is clear that tension between the Mising community and the Assamese community represented by *Satras* is not new, but based on certain philosophical and ritualistic differences. However, these differences have existed in Majuli for almost 500 years. Moreover, most of the Mising people in Majuli still accept the neo-Vaishnavite faith and adhere to the *Satras* as the chief social, cultural and religious institution of the island. However, riverbank erosion has the capacity to challenge the age-old power structure of Majuli. Riverbank erosion and the resultant loss of land has given new impetus to these latent differences and now there is a clear anger among Misings over their marginalization in both social and cultural life of Majuli as well as in disaster management efforts by government. The tension between communities that has always existed could easily become conflict if not resolved properly.

Coupled to this is the new foray of Christian missionaries in Majuli who have used the grievances of the Mising community against *Satras* to establish a sizeable missionary activity in Majuli. This has further complicated the situation as *Satras* are against any proselytizing in Majuli. To prevent this situation from becoming a more serious conflict, it is necessary that the unequal distribution of power and space and the impact of such distribution during and after disasters is clearly understood by policymakers, and proper disaster management plans are formulated and implemented that cater to the needs of all of the communities in Majuli.

References

Baruah, B. K. (1994). *Sankardeva-Vasihnav saint of Assam*. Guwahati: Bina Library.

Birkmann, J., & Wisner, B. (2006). *Measuring the unmeasurable: The challenge of vulnerability*. Bonn: UNU-EHS.

Blaikie, P., Cannon, T., Davis, I., & Wisner, B. (2014). *At risk: Natural hazards, people's vulnerability and disasters*. London: Routledge.

Bolin, R., & Bolton, P. (1986). *Race, religion, and ethnicity in disaster recovery* (Monograph No. 42). Boulder, CO: University of Colorado, Institute of Behavioral Science.

Chaliha, B. P. (1998). *Shankardeva studies in culture*. Nagaon: Srimanta Shankardeva Sangha.

Choudhury, R. N. (2011). Shankardeva's philosophy of religion. In G. Barua (Ed.), *Srimanta Shankardeva and his philosophy* (pp. 36–48). Nagaon: Srimanat Shankardeva Sangha Mills.

Collins, T. W. (2008). The political ecology of hazard vulnerability: Marginalization, facilitation and the production of differential risk to urban wildfires in Arizona's White Mountains. *Journal of Political Ecology*, *15*(1), 21–43.

Cuny, F. C. (1994). *Disasters and development*. Dallas, Texas: Intertect Press.

Das, D. (2014). 'Majuli in Peril': Challenging the received wisdom on flood control in Brahmaputra River Basin, Assam (1940–2000). *Water History*, *6*(2), 167–185.

Das, M. (2013). *Socio economic and demographic consequences of river bank erosion in majuli island Assam* [PhD Thesis] NEHU, Shillong. Retrieved from https://shodhganga.inflibnet.ac.in/handle/10603/169808

Haas, J. E., Kates, R. W., & Bowden, M. J. (1977). *Reconstruction following disaster*. Cambridge, Massachusetts: The Massachusetts Institute of Technology.

Kasperson, J. X., & Kasperson, R. E. (2005). *The social contours of risk: Risk analysis, corporations & the globalization of risk* (Vol. 2). London: Earthscan.

Kelman, I. (2003). Beyond disaster, beyond diplomacy. In I. Kelman (Ed.), *Natural disaster and development in a globalizing world* (pp. 126–139). London: Routledge.

Nath, D. (2009). *The Majuli Island: Society economy and culture*. Kolkata: Anshah Publishing House.

Oliver-Smith, A. (1990). Post-disaster housing reconstruction and social inequality: A challenge to policy and practice. *Disasters*, *14*(1), 7–19.

Oliver-Smith, A., & Hoffman, S. M. (1999). Anthropology and the angry earth: An overview. In A. Oliver-Smith & S. Hoffman (Eds.), *The angry earth: Disaster in anthropological perspective* (pp. 1–16). New York: Routledge.

Peacock, W. G., & Bates, F. L. (1982). *Ethnic differences in earthquake impact and recovery. Recovery, change and development: A longitudinal study of the 1976 Guatemalan Earthquake*. Athens, GA: The University of Georgia.

Pelling, M. (1999). The political ecology of flood hazard in urban Guyana. *Geoforum*, *30*(3), 249–261.

Quarantelli, E. L., & Dynes, R. R. (1976). Community conflict-its absence and its presence in natural disasters. *Mass Emergencies*, *1*(2), 139–152.

Robbins, P. (2004). *Political ecology*. Oxford: Blackwell Publishing.

Sahay, A., & Roy, N. (2017). Shrinking space and expanding population: Socioeconomic impacts of Majuli's changing geography. *Focus on Geography*, *60*, 1–26.

Sarmah, P. J. (2012). *Natural resources and livelihood pattern among the Mishing community of Majuli island* [PhD Thesis] NEHU, Shillong. Retrieved from https://shodhganga.inflibnet.ac.in/handle/10603/169792

Sen, A. K. (1981). *Poverty and famines, an essay on entitlements and deprivation.* Oxford: Oxford University Press.

Sorokin, P. A. (1946). *Man and society in calamity: The effects of war, revolution, famine, pestilence upon human mind, behavior, social organization and cultural life*. Piscataway, New Jersey: Transaction Publishers.

Zaman, A. (2015). Mortuary rite among the Mishing tribe in a rural context of Assam. *Cultural and Religious Studies*, David Publication, July–August, *3*(4), 177–184.

5 A Regional Political Ecology of the Changing City-Scape and the Crisis of Conservation of the East Kolkata Wetlands (EKW), India

Sana Huque

5.1 Introduction

Until recently, through the usage of the idea of 'sustainable development', the focus was predominantly upon addressing the natural, social and economic concerns of growth. However, the current world is such that despite quickly approaching the tipping point it is witnessing the increasing normalization of acts that go against the intent of safeguarding the environment and ecology of the planet. Actions of the US President's withdrawal from the Paris Agreement on climate change, the Brazilian President doing the barest minimum to control the raging fires and massive deforestation of the Amazon rainforests and the Indian Ministry of Environment and Forest proposing vast dilution of Environment Impact Assessment (EIA) procedures so as not to hinder businesses are among the multitude of similar instances that speak of the fact that politics and politics alone has become the driving economic force in the current world.

In the context of wetlands, perception about such areas and economic logic of the time have always attempted to give new meaning and identity to them. In the case of the East Kolkata Wetlands, it has been just that. As times have shifted from pre-colonial to colonial to the recent age of capitalist neoliberal development, the wetlands in the eastern vicinity of the city of Kolkata have morphed from disregarded to detestable to extremely desirable, based primarily on the perception and pecuniary rationale of the time. However, even though politics is integral in perception formation and financial logic, oftentimes it remains sub rosa. In this context, Bryant (1991: 164) says, 'Indeed, the political dimension typically is ignored, and ecology de-politicized'. Walker (2006: 385) writes that perhaps, 'The most fundamental role of political ecology is to question the oversimplifying and misleading conventional views of human-environment relations'. Therefore

DOI: 10.4324/9780367486433-5

the objective of this paper is to highlight the role of politics as the fulcrum of economic and ecological decisions set in the context of Kolkata and the East Kolkata Wetlands.

5.2 The Theoretical Perspectives

As a concept, political ecology was conceived out of the need to integrate the concepts of political economy and the growing apprehensions about the state of the ecology of the world. According to Peet and Watts (2002), the term 'political ecology' materialized in the 1970s as a response to the growing theoretical need for a concept that could integrate land use practice with the local–global political economy and the reaction to the growing politicization of the environment. However, the concept is a range of definitions. Within its ambit, the emphasis is not on political economy alone, the role formal political institutions play, environmental change and even narratives and stories about such changes are important arenas within the concept (Robbins, 2012: 14). Despite the varying emphasis, what never changes is the concept's firm grounding in being an explicit alternative to 'apolitical' ecology of the type of neo-Malthusian theories about population growth and environmental degradation (Walker, 2006).

In the context of this chapter, however, the political ecology framework that emerges useful is the one suggested by Blaikie and Brookfield in which environmental degradation is analysed from not just economic and ecological aspects but also a political lens. Blake and Brookfield define the concept in the following terms, 'The phrase "political ecology" combines the concerns of ecology and a broadly defined political economy. Together this encompasses the constantly shifting dialectic between society and land-based resources, and also within classes and groups within society itself' (Blaikie & Brookfield, 1987). The definition by Watts (2000: 257) says that concept of political ecology is an attempt 'to understand the complex relations between nature and society through a careful analysis of what one might call the forms of access and control over resources and their implications for environmental health and sustainable livelihoods'. This understanding of the concept is an acknowledgement of the environmental conflict emanating from struggles over knowledge, power and practice as well as politics, justice and governance (Robbins, 2012:16).

I have also added a historical dimension to give perspective to the politics of economy and environment as can be observed in the context of the wetlands of East Kolkata, for as Escobar (1999:3) says political ecology can also be 'the study of the manifold articulations of history and biology and the culture meditations through which such articulations are necessarily established'. Another definition of political ecology that is pertinent to this paper is by Stott and Sullivan (2000: 4) who called it the discipline which 'identified the political circumstances that forced people into activities which caused environmental degradation in the absence of alternative possibilities ... involved the query and reframing of accepted environmental narratives, particularly those directed via international environment and development discourses'. This definition is a clear elucidation of the 'political dimensions of environmental narratives ... to suggest that accepted ideas of degradation and deterioration many not be simple linear trends that tend to predominate' (Stott & Sullivan, 2000: 5).

Apart from political ecology, a concept like the political economy of urbanization, especially the one to emerge as a consequence of the post-industrial information era, lends an important theoretical basis to the observations that form the ambit of this chapter. Theories of spatial manifestations of capital investment and consumption in urban areas assist to understand how urban landscapes across the globe have rapidly transformed as a consequence of cities becoming 'strategic territories' (Sassen, 2005: 27) through which flow the global capital, labour, goods, raw material and people. Harvey's concept of capital circuits, a concept in the Marxist political economy, demonstrates the points of contact between economic processes, urban processes and politics. Harvey shows how reinvestment instruments of capital, accumulated over time from labour, can initiate and sustain rapid transformations of urban landscapes.

While Harvey speaks about capital's influence upon the built landscape, Smith's idea focuses on the spatial consequences of systems of production. Smith's theory says that development of cities is dependent upon location and land use policies that get adopted by corporate firms (Procter, 1982). The motive of profit maximization among corporate firms that leads to technological advancements in spheres of transportation, communication and production, as well as concentration of capital, also determines the type of land use that emerge in cities which is characterized by 'office park central cities, suburbanisation of industrial employment, and regional shift of industry to the sunbelt' (Procter, 1982). In the context of Kolkata the land use change as a consequence of capitalist development is easily discernible. Smith also says that, in the wake of corporate relocation and land use decisions, come capitalist enterprises who are geared towards land speculation and profitable development of built environment (Procter, 1982). Kolkata's changing wetlands allude to this observation of Smith. The role of the state in the spatial organization of cities by corporations act on both national as well as city levels. At the national level the state plays a supportive and enabling role for the location policies of the capitalist economies through provision of subsidies, economic support, even physical infrastructure. At the level of the cities, the state's role is to use its clout to further the economically determined urban form by posturing it as city planning.

As the capitalist development consolidates itself as a global phenomenon, the idea of land grabs has also emerged as an important discourse in political economy of urban development. In the context of East Kolkata Wetlands the phenomenon of land grab is particularly important. In India a number of instances of land grabbing have happened under the aegis of the state which is trying very hard to put the country on the global capital map. After India went the neoliberal way, especially post 1990s economic liberalization, there has been a rising conviction among the political class that,

> Following in the true tradition of the distributors of grace, global capital doles out the goodies to those who offer the best tributes in terms of tax exemptions, subsidized provision of natural resources like land and water, and the like.
>
> (Basu, 2007: 1281)

As a consequence the state has taken upon itself the role of facilitators for global capital to operate with ease in the country and find land for the purpose, most times by resorting to the colonial Land Acquisition Act of 1894.

The impact of such actions upon the cities have been their conversion into arenas of action where large investments, scams and people's resistance movements against evictions have become part of the urban restructuring process, along with the rising number of instances of land grab (Mahadevia, 2011). West Bengal itself, where the left had stormed to power based on their pro-poor and pro-farmer politics and had managed to hold onto it for nearly 40 years, adopted the contentious policy of land grabbing when it began the tread on the path of globalization in pursuit of development. In the context of Kolkata's neoliberalism induced urban restructuring and the resultant impacts upon the wetlands, examining the state's policy perceptions become a critical exercise because of its visibly increasing techno-economic optimism and administrative managerialism that usually surround the idea of capitalist growth of cities (Borras, Hall, Scoones, White & Wolford, 2011).

Land grabbing has emerged particularly as a challenge for areas that identified as 'marginal lands' or are perceived as 'idle', 'waste' or 'unoccupied' land and in most cases are usually the land in peri-urban locations. The East Kolkata Wetlands is often labelled as peri-urban of Kolkata. What is not acknowledged is that it is also a social, economic and environmental space where three systems, namely, agricultural system, urban system and natural resource system are in constant interaction. The idea of East Kolkata Wetlands as peri-urban can be understood as what Narain (2009) describes as 'a process or an analytical construct instead of just a place that allows the examination of relationships between rural and urban activities and institutions as well as the relationships between urban centres and villages'. Peri-urban areas, unfortunately, are also the spaces that continue to be at the receiving end of the ever expanding urban limits. Consequently land grabbing, displacement of agricultural communities and ecological damage can never be too far behind.

Like it is in the case of land grabbing, there exists a political economy of peri-urbanization which concerns with what structural factors are driving change, who benefits from this change, and how does this change affect the politics of the development processes? (Shatkin, 2016: 141) Shatkin's (2016) research shows that there can be two politico-economic perspectives to peri-urban development. The first sees it as a story of hope where the opening up of the peri-urban space for economic growth, through judicious state action, strategic infrastructure investment and economic planning, can usher significant economic benefits for the residents of the peri-urban area. The second perspective, however, focuses on the fear of peri-urban development turning into a chronicle of violence and dispossession at the hands of the government, it acting as destroyers of communities, environment and livelihoods in alliance with capital. What is also another outcome of peri-urban development is the escalation of land prices in such areas. In this context, Shatkin (2016 142) says,

> Across much of urban Asia, dramatic increases in land prices presents state actors with acute opportunities and challenges, leading them to develop new strategies aimed at tapping into real estate markets as a means to gain financial power and greater control over urban spatial change.

States throughout Asia have used the government's powers to monetize land by facilitating land value hikes. It has been done either by directly extracting revenue for government from land development, or by distributing the profits of land development to powerful corporate backers of the state (Shatkin, 2016). Either way, such means of land management usually have far-reaching impacts upon urban politics as well as the manner in which spatial patterns of urban development get determined. This also reveals important details about the emerging patterns of social and spatial disparities and political contentions. These tendencies also have a close association with Goldman's observations about the rise of 'speculative urbanism' or the increasing emphasis by state actors on the development of world class built environment which ends up fuelling corporate land speculation (Shatkin, 2016).

The impact of booming real estate and the ensuing developments mentioned above, especially in a city like Kolkata, has been immense upon the peri-urban areas. Aggressive efforts have been launched in the state to transfer and sell the peri-urban areas of Kolkata to industrialists and developers. Rajarhat and New Town are the most recent examples in this regard. While it is true that West Bengal saw massive opposition to land transfer attempts by the state to Special Economic Zones (SEZs) and industrialists from those who work and reside in these peri-urban areas, such as the farmers of Singur and Nandigram, the reality is that the land monetization of the peri-urban areas of Kolkata has mostly witnessed political and legal reforms and elite collective actions that push the envelope in favour of land acquisition efforts (Shatkin, 2016). In this context Kundu (2016: 98) quotes Ananya Roy, who argues,

> The reconfiguration of the urban periphery of Kolkata was closely aligned to the ways in which the upper- and middle-class elite of Kolkata wanted to assert their *bhodrolok* identity by demanding a world class township that would be integrated into the circuits of global capital and yet would bypass the poverty and squalor in the existing core city.

5.3 The (Spurned) Wetlands of East Kolkata

Ramsar Site No. 1208, more commonly referred to as the East Kolkata Wetlands, is an expanse of wetlands which is approximately 125 sq. km or 12,500 hectares in areal extent, sitting on the eastern periphery of Kolkata. To many residents of Kolkata the wetlands are actually a 'wasteland' where the city sends its garbage and sewage in order to get rid of them. According to a septuagenarian living in central Kolkata, 'Such a vast wasteland should not be allowed to lie vacant when the city requires its economy to pick up'. In many imaginations, while the wetlands are a wasteland, in so many others the area does not exist at all. 'Kolkata has wetlands! I've heard about Dhapa because that is where all our garbage goes. But I've never heard that there are wetlands there' is what was said by a school teacher. Others said, 'It is a place best avoided because except for garbage nothing else can be found there' and 'normal people from the city should never venture into such places'. For many in Kolkata the area east of the city has remained either a blindspot or an area about which 'not much is known because this place is completely cut-off from the city'. In reality EM Bypass, which is one of the busiest roads of Kolkata, sits right next to it and many popular residential areas like Ruby, Bengal Peerless apartment complexes and Ajay Nagar stand just adjoining the wetlands.

Making such associations with the wetlands is not new. Historical records left behind by the British colonizers are replete with such abject references for the wetlands. Alexander Hamilton's account from 1706, which is also valued as the earliest authentic English account of Calcutta and its vicinity, the marshlands that are precursor to the East Kolkata Wetlands are defined in the following terms:

> Mr [Job] Charnock chose the site of Calcutta for the sake of a large shady tree, tho' he could not have chosen a more unhealthy situation on all the line of the river [Hugli], for 3 miles to the east is a salt water lake which overflows in September and October and then prodigious number of fish resort there, but in November and December when the floods are dissipated, those fishes are left to die and, with their putrefaction, affect the air with the thick, stinking vapours which the north-east winds bring with them to Fort William, so that a great yearly mortality is caused by them.
>
> (Chattopadhyay, 1990: 6–7)

In James Ranald Martin's *Note on the medical topography of Calcutta* from 1837, the Salt Lakes are portrayed as unsanitary locations which should be reclaimed for the sake of the health of the European inhabitants (Chattopadhyay, 1990). The early settlers, therefore, worried not only about how to control the affairs of the marshland but actively contemplated upon how to subjugate and reclaim it. This marks the beginning of forgetting and repressing of the 'soaking ecologies' of the Bengal Delta (Bhattacharyya, 2018).

The post-partition period saw an acceleration in the dismantling of the wetlands. It was the period of setting up of big townships to house the rapidly growing population. In 1953, based on the investigation by the Drainage Enquiry Commission, the state government decided that parts of the wetlands have to be reclaimed for urban expansion. As a result, Salt Lake township or Bidhannagar, East Calcutta township and Baishnabghata Patuli township were conceived upon the wetlands. Later more wetlands were converted to accommodate the Eastern Metropolitan Bypass (EM Bypass). Therefore, by the time 1980s descended, which is also the time when the city fell into its economically depressed state, these major townships had already been conceived. At that point it became important to utilize the wetlands in such a manner so as to give a sense of economic confidence that Calcutta could still think of undertaking economically intensive ventures like setting up major satellite towns on land reclaimed from wetlands.

Incidentally, the 1980s are also the time when the wetlands were rediscovered for the natural wonders it performs. The person instrumental in bringing to the fore the various functions of the wetlands, especially its resource recovering capabilities was Dr Dhrubajyoti Ghosh, a veteran ecologist and conservationist who dedicated his life to the cause of conservation of East Kolkata Wetlands. In his quest to understand what happens to wastewater in a city without a single sewage treatment plant, he managed to discover how the wetlands contribute to the survival of Kolkata. He says,

> In 1981, serendipity took me to these wetlands in search of an answer a question set by the state government: how can wastewater from the city of Kolkata be reused. I reached the unknown wonder of the world that I named then and there: the East Kolkata Wetlands. I mapped it along with the villagers and the area of 12,500 hectares was identified. Today, this wetland is a Ramsar site, declared so in 2002.
>
> (Ghosh, 2017)

Dr Ghosh went on to uncover for the world the mechanisms of this natural sewage treatment plant and named the area the East Calcutta Wetlands. The century-old traditional knowledge that had mastered the technique of utilizing waste water for cultivation of fish and farm produce became a part of mainstream ecological conservation knowledge, courtesy of Dr Ghosh.

The uniqueness of the East Kolkata Wetlands lies in its capacity to assume the role of an important urban infrastructure despite being a natural area that manages not only the water-related issues of a major city like Kolkata but also provides the city with ecological, economic, social and even cultural benefits. More recently, the East Kolkata Wetlands have been called a desired model for optimum reuse of water to combat the global water crisis and named among world's only two most precious natural resources for wastewater treatment by the secretary-general of Ramsar Convention, Martha Rojas-Urrego.

In addition, the East Kolkata Wetlands have been the location that sustains Kolkata's food security by producing nearly 40% of the vegetables sold daily and over 10,000 to 15,000 tonnes of fish supplied yearly to Kolkata's markets. The wetlands of Kolkata have not just been acting as providers of food to the city. It has been 'ecologically subsidising' the city by rendering several additional services that contribute to Kolkata's liveability. Dr. Dhrubajyoti Ghosh termed the ecological services that the wetlands provide as 'keystone services'. The manner in which a keystone species is known to maintain the structure and function of an ecosystem, similarly the wetlands through a harmonious collaboration between wastewater, sunshine, algae, fishes and itself produces keystone services that uphold and maintain the structure, integrity and biodiversity of the landscape. If these services are to go extinct then the consequences would be 'altered disturbance regimes, new patch dynamic equilibria and loss of integrity and/or dominant species' (Ghosh, 2005).

It must be noted that the results from the keystone services transcend not just the physical boundaries of the wetlands but in some cases also that of Kolkata. The locally beneficial keystone services emanating from the wetlands have been identified as an important repository of a wide variety of flora and fauna. Therefore the wetlands have been called multidimensional – an ideal urban metabolism model, area complying with fish criteria and also being an important waterfowl habitat (Kundu, Pal & Saha, 2008). Moreover, the area is an important site for migratory birds and many insects, mammals, reptiles, etc. The area also has a high diversity of microphones and phytoplanktons.

The East Kolkata Wetlands are a crucial flood control mechanism that by soaking up the flood waters keep the city from getting inundated. A study by Kolkata-based civil society organization South Asian Forum for Environment (SAFE) and International Water Management Institute (IWMI), Colombo found that only 48 hours of torrential is needed to completely submerge the city and it could happen as soon as 2020 if the wetlands are not restored to their previous capacity of assimilating flood waters. In addition, the wetlands recharge the city's groundwater, that have begun to dwindle due to excessive extraction. If the East Kolkata Wetlands are to go extinct the city could lose all its major groundwater recharging points pushing it towards subsidence and greater flood risks. The wetlands, although rarely attributed for these services, also provide stability to the sub-continental watershed and is a very important buffer for the Sundarbans and can potentially provide solutions that can ally the deterioration of the Ganga (Banerjee, 2012).

Yet the predominant narrative that continues to get propagated about the East Kolkata Wetlands is that the area serves no purpose of its own and hence the city can make unchecked ingress into it to serve the purpose of rapid urbanization of Kolkata. History, along with economics and ecology, say that the wetlands have always had a very close association with the city's evolution. Yet the area is hardly ever accredited in the chronicles of its evolution. This marshy stretch has had a long history, one can say even longer than Kolkata itself. From the moment of its inception to this very point in time the city continues to grow by shrinking floodplains and disappearing wetlands (Bhattacharyya, 2018).

In the project of dismantling the wetlands, the state, it's agencies as well as the citizenry have played a role. Almost every day in Kolkata now begins with some news outlining the extent of deterioration happening in the East Kolkata Wetlands. I have presented a short list of headlines from not too a distant past:

- Another bheri in Kolkata disappears (published in Times of India, 22 Feb, 2017);
- Mayor wants flyover through wetlands (published in Times of India, 17 May, 2017);
- West Bengal may lift building limits in fragile wetlands (published in Hindustan Times, 13 Dec, 2017);
- Greed and indifference are destroying east Kolkata's wetlands (published in Scroll.in, 7 April, 2016);
- Green Tribunal Raises Red Flag over Kolkata Wetlands (published in The Wire, 27 July, 2016)

An article appeared in 2012, title, 'Wetlands being sold in pieces' (Ray, 2012). According to it per *katha*[1] of the wetland in Choubagha is being sold at 3.25 lakhs. The news team came across labour contractor whose men were engaged in the construction of a wall around 1,040 *katha* of wetlands that happens to be located within the boundaries of the protected Ramsar site. The article quotes the contractor saying, 'The promoters want to sell off the land in 3 to 7 *katha* plots. Those who build houses here will be in the lap of nature' (Ray, 2012). When asked how he could be so sure that new buildings will not come up on the adjoining wetlands by similarly encroaching waterbodies, the contractor states, 'True, there is no guaranteeing that. Dozens of waterbodies have been encroached over the past few years and land sharks have been more active than ever. Anyway, it does not concern us as long as we get to earn' (Ray, 2012). The contractor further states that large parts of the adjoining *bheries* would definitely get grabbed by land sharks because of the 24 feet wide road expected to come up soon in order to woo more buyers to the area.

A report prepared by Society for Creative Opportunities and Participatory Ecosystems (SCOPE) in 2017 and titled *Not a Single Bill Board: The Shifting Priority in Land Use within the Protected Wetlands to the East Of Kolkata*, found that Bhagabanpur mouza, which used to have most extended stretches of water bodies in the wetlands, has transformed into the most densely populated area of the wetlands. The report says, 'The conversion of land has been formidable, with waterbodies changing from 88 per cent in 2002 to 19 per cent in 2016 and corresponding change in full settlement area from 0.18 per cent to 13 per cent in 2016' (Gupta, Chaudhuri, Ghosh & Sarkar, 2017). SCOPE's research found that along with the water bodies what has also become rare in the area is the unique knowledge of sewage purification through fish growing, spatial planning, analytical skills, approaches and methods. This has been the ramification of demographic change the area has experienced due to people coming and settling down from outside the wetlands with no idea of how the wetland functions

(Gupta et al., 2017). The title of the report, interestingly, alludes to the fact about how the periphery of the wetlands that sits along the EM Bypass has been inundated with advertisement billboards, yet not a single one can be found that indicates that East Kolkata Wetlands are a globally recognized natural marvel.

While flagging these violations, the question that arises is what actions have the state taken to stem the destruction of the wetlands. To search for answers the explanation that I chanced upon is by East Kolkata Wetland Management Authority (EKWMA), an agency constituted by the East Kolkata Wetlands (Conservation and Management) Act, 2006. According to EKWMA what explains the apathy is that complaints related to the wetlands are never of high priority. The following section therefore is a critique of the EKWMA. Based on the composition and function of the statutory body I attempt to examine if the EKWMA has managed to do what it set out to achieve and if not why.

5.4 The Role of East Kolkata Wetlands Management Authority: A Critique

The East Kolkata Wetlands Management Authority (EKWMA), consti-tuted under Section 3 of the East Kolkata Wetlands (Conservation and Management) Act, 2006, was brought into existence with the intention 'to provide for conservation and management of the East Kolkata Wetlands and for matters connected therewith and incidental thereto' ('West Bengal Act VII', 2006). Although the wetlands have always suffered casualties from conversion and encroachment since the time Kolkata began its growth and transformation into a major metropolis, the setting up of the EKWMA can be considered as an acknowledgement of the social, ecological and eco-nomic significance of the wetlands and its status as a Ramsar site of inter-national importance.

However observations indicate that the creation of this statutory body has not been very effective in protecting the wetlands. In fact Banerjee (2012) observes that post the creation of EKWMA a paradoxical situation emerged in which the destruction of the wetlands accelerated instead of showing a decline. What needs to be questioned then is has the EKWMA been conceived in a manner to be really effective against wetlands destruc-tion? If not then what are the flaws that are inherent to this statutory body?

A read through the 2006 Act gives hints as to how the composition and functions of EKWMA has been conceived. The list of members is a lengthy one – 19 members in total. However, in this long list only four members can be distinguished as persons with expert knowledge about wetlands. The Act names them as (a) representative of the non-government organizations having expertise in the field of wetland conservation; (b) representative of the non-government organizations having expertise in the field of wetland management; (c) representative of the fishermen's co-operative societies formed under the West Bengal Inland Fisheries Act, 1984; and (d) representative of the Institute of Environmental Studies and Wetland Management, Kolkata. All four experts are to be nominated by the State Government and have not been bestowed with any real powers. The Act expects them to be involved in advisory capacities only.

The remaining majority of the members in EKWMA are secretaries from various departments of the state government like environment, urban development, municipal affairs, land and land reforms, irrigation and waterways, fisheries, panchayat and rural development, district magistrates of North and South 24 Parganas, commissioner of Kolkata Municipal Corporation, member-secretary of West Bengal Pollution Control Board, and chief executive officer of Kolkata Metropolitan Development Authority. The composition of EKWMA indicates that it is largely a bureaucratic body and hardly an expert body for wetland management. The various departments may have the best interest for the wetlands in mind but it does not clarify as to why a body that is aiming for conservation of wetlands does not have more people with specific expertise about wetlands at its helm. No explicit mention of incorporating the expertise of wetland scientists is mentioned anywhere in the act (Banerjee, 2012).

The question thus arises is whether the EKWMA is an embodiment of the new style of governance that has emerged in the country, namely managerial governance, which is thought to be 'more cost-conscious, efficient and effective paradigm of rendering public service' (Banerjee, 2012: 106). Since the time the nation began its walk on the path set down by the new economic policy, the government has become more managerial in the conduct of its affairs. By interpreting the problems of the wetlands through the lens of management, the state government has created an impression that it is possible to ensure the survival of the fragile ecosystem by merely managing the problems as they emerge from time to time. Banerjee (2012: 106) writes,

> From this managerial perspective—that everything including anthropogenic imbalances in natural systems can be technologically fixed—the scientists' in-depth knowledge of the complex dynamics of an ecosystem is not deemed necessary, as is evident from the bureaucratic nature of the management committee of the EKWMA.

The major functions that have been entrusted with EKWMA range from determining the boundary of the protected site to maintaining the area's wetlands characteristics to promoting its overall conservation. But instead of demarcating its boundary the EKWMA has modified the areal spread of the wetlands in recent years irrationally without any deliberations and discussions with either the wetland community or the experts. When ground realities indicate a severe decline of wetland area because of indiscriminate conversions, on paper there have been attempts to show an increase in wetlands area. The EKWMA has added 4 mouzas which are predominantly urban and not wetland, namely Nonadanga (Tiljala PS), Kochpukur (Bhangor PS), Kalikapur (East Jadavpur PS) and Thakdari (Rajarhat PS), to the original number of 32 mouzas. As per Dr Dhrubajyoti Ghosh this action has been taken by the EKWMA without any deliberations with the environmentalists who have been invested in the wetland's conservation. So although the current area under wetlands on paper is 12,741 hectares in reality the area may have declined from 12,500 hectares which was originally indicated in the map approved by Ramsar authorities in 2002. Other functions of EKWMA which look into prevention of illegal constructions and conversions have also been failures so far on its part. In fact it has been brought to the attention of the Ramsar secretariat that the state government has been gearing up to legalize the illegal constructions within the wetland area by modifying the rules of conservation.

In 2017, the 2006 Act was amended. According to the East Kolkata Wetlands (Conservation and Management) (Amendment) Act, 2017, the EKWMA was reconstituted to anoint the State Environment Minister Sovan Chatterjee[2] as the new chairman of EKWMA, replacing the Chief Secretary of the state. Such a move is clearly an indication of politics creeping into the bureaucratic set-up of EKWMA. On this Biswajit Mukherjee, the former chief law officer of the state pollution control board, said, 'The Supreme Court has made it clear that political executives should not head environmental boards or authorities. A minister becoming the chairman of the East Calcutta Wetlands Authority is a violation of the apex court's order in spirit' ('Sovan in charge of wetlands', 2017).

Therefore, further dilution of the Act can be gleaned from the manner in which the composition of EKWMA has been modified. Currently operating with only 13 members, the Authority no longer accommodates any representatives from the State's Pollution Control Board or members from the Institute of Environmental Studies and Wetland Management. On the contrary, the additions include a high level official from the Department of Tourism and four nominated 'experts' from areas of 'wetland ecology, hydrology, fisheries and socio-economics' ('West Bengal Act V', 2017). With no room for representation of civil society groups or wetland community members, the new act almost explicitly shuts doors on those with real interest and expertise in the field of wetland conservation.

What further complicated the stance of this body is the fact that the then ex-officio chairman of EKWMA was also the minister in charge of Housing Department of West Bengal Government at the same time. This conflict of interest probably explains the erstwhile state environment minister's views about the wetlands and his plans for its future. A few months prior to nominating himself as the EKWMA chairman, the minister was in news because of his statements which are not only anti-environment in general and but also anti-wetland in particular. After taking charge he immediately began a campaign against environmental safeguards and specifically stated that the protected status of the wetlands need to be done away with to clear the way for developmental projects in the area. He has said on record that, 'One cannot ignore the legitimate requirements of a city and blindly follow environmental restrictions because someone decided to go to an international body with the wetland map and get a huge chunk of land demarcated for protection' (Lokgariwar & Dewani, 2016).

Without doubt it can be said that the state's stand with respect to the wetlands have been both duplicitous and negligent. From the stance the state has adopted over the years point to how the state has taken a misleading stand when it comes to protecting the wetlands. For instance the wetlands received its recognition as one of a kind natural resource recovery system and water treatment facility in 2002 and the wetland authority came into existence in 2006. Yet even more than a decade later the state has still not shared or prepared a wetland management plan with the Ramsar authorities. In addition the state is yet to take any initiative to demarcate the area properly. Both these lapses show that even mandatory pre-conditions that the state had to fulfil upon the bestowing of the international recognition upon the East Kolkata Wetlands has not been carried out. The state is also behind on the submission of utilization certificates for the funds it received because of the East Kolkata Wetlands since 2012. The same environmental minister of Kolkata even proposed for dilution of restriction in order to permit prohibited activities in the area. The following are the proposals that the minister has pitched with respect to the East Kolkata Wetlands:

- Modify the existing conservation plan that forbids changing the natural character of the area to accommodate a garbage dumping ground within the confines of the wetlands.
- Bestow mutation rights on the people who live there. According to environmentalists if allowed the rate of wetland conversion is only expected to take a huge leap as it will attract a greater influx of those wanting to own residential land closer to prime city location at cheaper rates.

- Support proposals of road and flyover construction that will cut through the wetlands. Currently three roads have been proposed that will be built to pass through the wetlands. The first one is the elevated road that wants to connect Sector V in Salt Lake with Bantala. The road is the brainchild of Nabadiganta Industrial Township Authority (NDITA) that functions under the municipal affairs department. The areas to be affected like Natar Bheri, Goltala, Chakher Bheri and Dhapa are all situated within the wetland complex. The second is a proposed flyover to connect the Maa flyover with Rajarhat. Ironically the mayor selected a function to mark the World Environment Day to announce this 6.5 km flyover through wetland area. Reiterating his development vs environment justification, the mayor said,

> If we can build one of the longest flyovers in the city which will reduce the time of commuting to a great extent, where's the harm. We are not destroying the entire wetlands. We need a small part of it for construction of piers without disturbing its existence.

A third proposal that is doing the rounds is to build a 600 crore flyover to connect the airport with Science City. As per estimates 10–12 *katha* of the wetlands will be destroyed. The government has promised a compensatory digging of a 25-*katha* lake in the adjacent area ('East Kolkata Wetlands Under Threat', 2017), but it does not take away from the fact that the wetlands will still stand affected.

- Establish a zoo and an amusement park in the wetland area. Envisioned on lines of Singapore's Jurong Bird Park and Bangkok's Safari World the site has already been identified and survey of the area undertaken.
- Push to 'properly utilise' the wetlands as the environment minister sees them as a vast stretch of land that is lying vacant. This is being termed by the environmentalist of the city as an attempt to legalize the illegal structures that have come to crowd the encroached areas.

The above points list how the wetlands have become a target for destruction in the name of development. While the government is pushing its agenda of developing the wetland area for infrastructure, amusement and housing needs of the city, in the same breathe it talks about conserving the wetlands. The State Environment Department has proposed the creation of a 'wise use plan' for the wetlands. However it remains to be seen if the state's definition of wise use corresponds with the wise use definition that Ramsar explicates, i.e., 'The maintenance of their ecological character, achieved through the implementation of ecosystem approaches, within the context of sustainable development' ('Bengal to frame "wise use" plan', 2017). The state government has also declared that it will create groups who will be exclusively entrusted with the monitoring and prevention of waterbody filling and conversion in the wetland area. The groups are being called the *Jolabhumi Raksa Dal* or Wetland Protection Group. The proposed task force will comprise the police, representatives from state administration ranging from the level of state, district to the block level and environmentalists. Apart from monitoring and prevention the group shall also be responsible for returning the violated waterbody to its previous condition. However, since its formation not much has been reported about the activities of these groups.

The environment minister's dig at those who have been striving to protect the wetlands from the relentless onslaught of urban expansion exposes how little he knows about what it means to have a Ramsar site adjoining a city. Statements like 'wetlands conservation mean little to the common man' and managing to get away from widespread public condemnation show how little the city cares for such a rich ecological asset. This silence gives away the reality of most citizens not even being aware of a vast stretch of wetlands lying next to the east of Kolkata performing the job of sewage management, waste recycling, flood control, providing food security and delivering of several additional ecological services for the city. To say that the wetlands mean little for the common man or the city at large is a statement made out of utter lack of knowledge about the richness of the wetland ecosystem and how it contributes towards the survival of the city. Under such circumstances then one is bound to ask, where is the outrage, why are the citizens not protesting the irreversible damages to the wetlands or is the mayor right about the common man not being concerned about the environment? The following section attempts to explore the reasons that may be behind the out of the ordinary silence on part of the civil society of the city in the face of vandalism of an internationally recognized natural asset like the East Kolkata Wetlands.

5.5 The Middle-Class Inducers of East Kolkata Wetlands Transformation

An attempt has been made to look at the wetlands falling victim to encroachment and conversion from the perspective of it being a class problem, particularly the middle class. The hegemonic influence of the middle class upon Kolkata is visible in its institutions, in the city's culture and most prominently in the political discourse of the city. The middle class has emerged as such a seminal group in itself that even the country apparently has to depend on it for its growth. The middle-class' desire and ability to pay a little more for a better quality of life can evidently create and nurture markets and encourage economic productivity. This is why Das (2000) says, 'The most striking feature of contemporary India is the rise of a confident new middle class ... whether India can deliver the goods depends a great deal on it'. Such a perception of the urban middle class, says Fernandes (2006), points to its growing significance in local, national and transnational imaginations in globalization of the twenty-first century.

In Kolkata, The term middle class was designated for those who were associated with government jobs or with other professional services until the 1990s. These salaried elites came to wield great influence upon the Bengali society because of their close association with the state sector. This class worked tirelessly to involve itself with the institutions and the politics of the state creating a cult of self-assured bourgeois nationalism (Donner & Neve 2011: 4). However, since the 1990s a new middle class has emerged. This new middle class has characteristics that are more close to the 'nomenklatura capitalist' of Russia than the middle class of the Nehruvian era. Because of neoliberalism the middle class now enjoys not only the privilege of having close relations with the political institutions of the state but also the benefit of being economically well-off. Donner (2015: 130) writes,

These transformations are particularly pertinent in urban areas, where ... there is nothing new about the middle class as an important player in politics, but where the so-called new middle class stake their claim to status more exclusively on economic standing, the outward sign of which is participation in the rituals of consumerism evidenced and experienced within the new spatial order of cities across India.

The impact on the wetlands of such privileged outlook of the middle class and upper middle class has been the attempt to co-opt the wetlands for housing and recreation for themselves while leaving the original inhabitants of the area deprived of their home and livelihoods. The affluence that is visible along the stretch of EM Bypass adjoining the wetlands shows the direction in which the state of West Bengal wants to move towards based on the aspirations of the privileged class. So as a frontier of the expanding city the area has rapidly converted into a space for the *nouveau riche*. I take this exhibition of development, despite it threatening the wetlands, as a display of how environment is getting re-configured as a middle-class idea and the silence of the citizens as being representative of their interest in availing the benefits of neoliberal development.

So how does the wetlands appear in the middle-class imagination? Like the British, the wetlands are still looked upon as land-in-waiting by the middle class. What began as an endeavour to reclaim the wetlands to provide housing facilities to mostly the lower income families has in the present scenario metamorphosed into creating spaces for homes, workspaces, and entertainment centres for the affluent middle class of the city. The housing complexes, the government and private offices, and the shopping centres in Bidhannagar openly allude to this point. In fact the Bidhannagar project was deemed so successful that the focus soon shifted to creating wider roads and parks every 3 kms incorporating the *bheris* of the wetlands so that an image of a holistic urban system emerges that appears to be focused on elevating the quality of life (Bose, 2015: 96). The later projects like the Baishnabghata Patuli Township and the East Kolkata Township, and the more recent Rajarhat New Town Project have all followed in the footsteps of Bidhannagar catering more to an affluent vision of a city than go for a more inclusive plan that also accommodates the economically weak (Huque, Pattanaik & Parthasarathy, 2020).

In the history of Kolkata's formation, while glorifying these developmental initiatives what is often forgotten is that the wetlands and its people have lost thousands of hectares of *bheris* and agricultural land because of it. The extent of loss can be imagined if the fact that Bidhannagar alone claimed nearly 4,000 hectares of the wetlands is taken into consideration. Those at the receiving end of the reclamation projects have not only lost their lands and waterbodies, but in some cases they have had to face violent eviction drives, police atrocities and even illegal coercive measures to make them vacate their lands. Therefore, livelihood concerns of the people dependent upon the *bheris* and agricultural land were also severely disrupted because of these developmental projects. For instance, prior to 1995, Rajarhat used to be a fertile agricultural area interspersed with villages with long histories of settlement, orchards, flower nurseries, ponds and substantial waterbodies (Kundu, 2017). Inhabited by farmers and fishermen, two-thirds of whom were from either Muslim or Dalit communities, the locality was connected to the city through the exchange of fresh produce, goods, and services (Dey, Samaddar & Sen, 2013). The villages of Rajarhat had electricity, functional primary schools, access to water, roads and irrigation facilities (Kundu 2017). But all of this was disrupted to accommodate the large-scale project. Over 1,00,000 persons became landless and jobless (Sengupta, 2013).

Several studies are available about the changes in number of persons practising agriculture in the area post Rajarhat development and all the estimates show a decline. As per Census figures for rural area the number of cultivators in 2001 were 4,261 and agricultural labourers were 7,217 (Karmakar 2015: 136). The figure for 2011 shows a marked declined to 2,798 for cultivators and 2,473 for agricultural labourers. While the census figures show a definite decline, some studies have produced numbers that reflect the alarming nature of livelihood loss in Rajarhat. The study by Chattopadhyay and Majumdar (2002) estimates that around 100,000 persons lost their livelihoods. Another estimate by Dey et al. (2013: 37) claims the figure to be 131,000. The figures compiled from newspapers and rights protection organizations are as follows – as of September 2002, 6,170 marginal farmers; 2,105 small farmers; 4,605 landless labourers; 4,000 fisherfolk; and 2,000 other families lost their livelihoods because of the township project (Dey et al., 2013: 37). Because of these changes the livelihood options in the area has changed and non-agricultural jobs have increased by 20 percent (Karmakar, 2015: 136).

Even without looking at the figures, a glance at the satellite images of Rajarhat from when reclamation and conversion for the township began and at present is enough to indicate the massive nature of land use change the area has experienced. The following satellite images are a comparison of one of Rajarhat's Action Areas from 2005 and 2019. The images clearly indicate that in just over a decade the area has been completely urbanized.

Currently a drive down the six-lane road through the heart of Rajarhat reveals a picture of a planned township with vast open spaces adorned with modern art installations, international convention centres and an art museum. Most people who are now either living in Rajarhat today or are investing in a flat in the area have nearly zero idea about the tragedies that have unfolded there. A massive eco-park called Prakriti Tirtha spread over 190 hectares runs along the Rajarhat Bypass for a considerable stretch. The park not only boasts of having very believable replicas of the wonders of the world but also has 'ecological zones' where one can come to experience wetlands, urban forests and grasslands. As per the website of the eco-park it is a 'living classroom about nature's service' where while 'relaxing, rejuvenating, socialising or seeking solitude' one can learn about how the wetlands filter water and plants filter air. However like ubiquitous ecological zones, the park also boasts of a number of ticketing booths sprinkled generously inside and outside the complex. In addition Housing Infrastructure Development Corporation (HIDCO), a state infrastructure development agency, has recently said that it has initiated India's

Table 5.1 Cultivators and Agriculture Labourers in Rajarhat in 1981, 2001 and 2011

Type of cultivation	Urban area			Rural area		
	1981	2001	2011	1981	2001	2011
Cultivators	126	580	327	5,164	4,261	2,798
Agriculture Labourers	577	326	349	6,531	7,217	2,473

Source: Census data 1981, 1991 and 2001 as cited in Karmakar, 2015.

(a)   (b)

Figure 5.1 Rajarhat/New Town Area in 2005 and 2019. The satellite image on the left shows the prevalent nature of land use in 2005 in Rajarhat, that eventually became New Town. The satellite image on the right shows the present form of land use. The vast open stretches with wetlands, agriculture fields and fisheries are now a thing of the past, obliterated by the urban landscape of New Town. Source: Google Earth, 2020

first man-made urban forest project within the eco-park premises. As per HIDCO chairman,

> An 'Urban Forest' is an area of large trees in a city setting. In Eco Park of New Town, an area of 5 acres near Gate 3 is nurtured as a Tropical Rain Forest. There are other zones in the Eco Park outside it. But the Tropical Rain Forest with its unique blend of evergreen trees, birds, mist creator and wild animal statues is certainly the most unique in the country.
>
> (Biswas, 2018)

Sadly, the irony is lost on both the HIDCO chairman as well as the people of the city who are flocking to the eco-park making it one of Kolkata's biggest attractions. No one cares to contemplate upon the fact that prior to the township the area was performing the same natural activities on a more extensive scale and instead of statues of animals the area had real animal and plant life. The area was also not behind a paywall that determines who gets to enjoy the pleasures of the ecological zones and who does not.

What is to be noted about Kolkata currently is that the Bidhannagar and Rajarhat model of development have become the templates to follow in the development of the new areas of Kolkata. Consequently similar motifs of development have become a common sight along the entire stretch of Biswa Bangla Sarani which has now subsumed the EM Bypass. Even the protected Ramsar site has become another area for a park in the eyes of the state and the people. Now there are plans to develop an East Kolkata Wetlands Park with a Nature and Wetland Interpretation Centre, a zoo and a bird park within the East Kolkata Wetlands complex. In this context Bose poses very pertinent questions whose answer the state needs to look for before it sets out to reclaim and convert more wetlands. He asks,

> So for whom is this initiative designed? For the ... people that already live and work in the Wetlands? The working poor in the city who rely on the foodstuffs produced in the wetlands fisheries and farms for part of their sustenance? What will this centre achieve for them (the people who already live and work there)? What will it teach them about the environment in which they already live? Or will it serve instead as another 'edutainment' facility, complementing the other attractions and theme parks designed to turn the wetlands from a working space and lived ecosystem into a park and playground for those with a 'developed' environmental consciousness?
>
> (Bose, 2013: 145)

5.6   Conclusion

With the adoption of neoliberalism-inspired economic policies and declining emphasis on industrialization one may say that theoretically Kolkata is in the phase of second or reflexive modernity, a phase in which a society emerges as a risk society. This is the phase when society begins to obsess about its future safety from hazards and insecurities induced and introduced by modernization (Beck, 1992). However, in Kolkata currently the obsession is over matters that appear to bring it economic gains in the coming years.

Although the city has moved into a phase of modernity marked by globalization and the supposed end of tradition, there is a stark absence of initiatives to address the environmental dangers that will arise from the extinction of the wetlands. Terms like 'sustainable development' that are considered a product of reflexive modernity can be heard commonly in association with developmental projects that are currently underway in the city. But ecological modernization of institutions in which their economic practices respect ecological limits remain unfulfilled. Apart from a few eco-parks and an urban forest that are supposed to contribute to the sustainable development of the city, the essence of ecological modernization in economic practices remains noticeably absent in Kolkata.

Globally 'eco-alarmism' has been on the rise since the early seventies. But it became a surge only after the eighties, which saw the release of the Brundtland Report. Till then environmental ideologies had occupied the peripherals of social context. Post Brundtland Report they moved to occupy a central position in institutional organizations of modern society (Mol & Spaargaren, 1993). Since then ecological concerns and environmental awareness have progressed further to capture not only political and socio-cultural imaginations but also economic enterprises.

In India, because of it being a developing nation, this consciousness may have begun to creep in much later. Indira Gandhi's speech at the Stockholm in 1972 made it clear that, for developing countries like India, development and addressing the receipt of basic needs of the people are as important priorities as the concerns for the environment. In the case of Kolkata it can be said that the city grew an environmental awareness even later than the rest of the country, because not only was the state inward-looking, its biggest concern was its declining economic condition. However, modernity has finally caught up with the city and many ventures in Kolkata, such as real estate industry, corporates, hospitality industry, IT sector and so on, can be heard spouting ecological terms like 'sustainable development', 'green practice' and 'eco-friendly'. The citizens are also aware of the earth heating up and the glaciers melting. So awareness about major environmental problems does exist in the city. However, this awareness is not reflected in the words and actions of the administration or the citizenry.

So it seems that the city may have become a case for Giddens' Paradox. Although Giddens described the idea in context of climate change, the issue of wetland loss can also be examined through the perspective of this paradox. Giddens (2009: 2) said,

> Since the dangers posed by global warming aren't tangible, immediate or visible, many will sit on their hands and do nothing of a concrete nature about them. Yet waiting until they become visible and acute before being stirred to serious action will, by definition, be too late.

The city of Kolkata is similarly choosing to do nothing about the indiscriminate wetlands destruction. Even after innumerable newspaper articles and research projects the city chooses to wait for visible signs of repercussions of the damage to emerge.

Notes

1 A unit of measurement of land in which 1 *katha* is equal to 720 sq. feet.
2 Since June, 2018 Sovan Chatterjee is no longer the environment minister. He has been succeeded by Suvendu.
    Adhikari, and the Prof. (Dr.) Saumen Kumar Mahapatra as the minister-in-charge of the environment department of the state.

References

Banerjee, S. (2012). The march of the mega-city: Governance in West Bengal and the wetlands to the east of Kolkata. *South Asia Chronicle*, *2*, 93–118.

Basu, P. K. (2007). Political economy of land grab. *Economic and Political Weekly*, *42*(14), 1281–1287.

Beck, U. (1992). *Risk society: Towards a new modernity*. New Delhi: Sage.

Bengal to frame 'wise use' plan for East Kolkata wetlands. (2017, March 4). *Business Standard*. Retrieved June 8, 2017, from http://www.business-standard.com/article/news-ians/bengal-toframe-wise-use-plan-for-east-kolkata-wetlands-117030400633_1.html.

Bhattacharyya, D. (2018). *Empire and ecology in the Bengal Delta: The making of Calcutta*. Cambridge: Cambridge University Press.

Biswas, L. D. (2018, July 10). In the heart of concrete Kolkata, a man-made forest is emerging. *The Quint*. Retrieved from https://www.thequint.com/news/environment/man-made-greenforest-in-kolkata-rajarhat-urban-hub.

Blaikie, P., & Brookfield, H. C. (1987). *Land degradation and society*. London: Methuen.

Borras Jr., S., Hall, R., Scoones, I., White, B., & Wolford, W. (2011). Towards a better understanding of global land grabbing: An editorial introduction. *Journal of Peasant Studies*, *38*(2), 209–216.

Bose, P. S. (2013). Bourgeois environmentalism, leftist development and neoliberal urbanism in the city of joy. In T. R. Samara, S. He, & G. Chen (Eds.), *Locating right to the city in the global south* (pp. 127–151). Oxon: Routledge.

Bose, P. S. (2015). *Urban development of India: Global Indians in the remaking of Kolkata*. Oxon: Routledge.

Bryant, R. L. (1991). Putting politics first: The political ecology of sustainable development. *Global Ecology and Biogeography Letters*, *1*(6), 164–166. https://doi.org/10.2307/2997621.

Chattopadhyay, H. (1990). *From marshes to township east of Calcutta: A tale of salt water lake and salt lake township*. Kolkata: K. P. Bagchi.

Chattopadhyay, K., & Majumdar, K. (2002). Economics of environmental degradation: The case of East Kolkata wetlands. In M. Mukherjee & K. Chattopadhyay (Eds.), *Kolkata: The city of wetlands, uncared resource, unrecognised beauty and unexplained truth*. Kolkata: Department of Fisheries, Government of West Bengal.

Das, G. (2000). *India unbound: The social and economic revolution from independence to the global information age*. New York: Anchor Books.

Dey, I., Samaddar, R., & Sen, S. K. (2013). *Beyond Kolkata: Rajarhat and the dystopia of urban imagination*. New Delhi: Routledge India.

Donner, H. (2015). Whose city is it anyway? Middle class imagination and urban restructuring in twenty-first century Kolkata. *New Perspectives on Turkey*, *46*, 129–155.

Donner, H., & Neve, G. D. (2011). Introduction. In H. Donner (Ed.), *Being middle-class in India: A way of life* (pp. 1–22). Oxon: Routledge.

Escobar, A. (1999). After nature: Steps to an antiessentialist political ecology. *Current Anthropology*, *40*(1), 1–30. https://doi.org/10.1086/515799.

Fernandes, L. (2006). *India's new middle class*. Minneapolis, MN: University of Minnesota Press.

Ghosh, D. (2005). *Ecology and traditional wetland practice lessons from wastewater utilisation in the East Calcutta wetlands*. Kolkata: Worldview.

Ghosh, D. (2017, October 22). Dispensation of a failed ecologist. *The Statesman*. Retrieved from http://www.thestatesman.com/opinion/dispensation-failed-ecologist-1502515047.html.

Giddens, A. (2009). *Politics of climate change*. Cambridge: Polity.

Gupta, D. D., Chaudhuri, S., Ghosh, S., & Sarkar, S. (2017). *Not a single bill board: The shifting priority in land use within the protected wetlands to the east of Kolkata*. Retrieved from https://www.scopekolkata.org/summary/.

Huque, S., Pattanaik, S., & Parthasarathy, D. (2020). Cityscape Transformation and the Temporal Metamorphosis of East Kolkata Wetlands: A Political Ecology Perspective. *Sociological Bulletin*, *69*(1), 95–112.

Karmakar, J. (2015). Encountering the reality of the planning process in peri urban areas of Kolkata: Case study of Rajarhat. *Archives of Applied Science Research, 7*(5), 129–138.

Kundu, N., Pal, M., & Saha, S. (2008). East Kolkata wetlands: A resource recovery system through productive activities. In M. Sengupta & R. Dalwani (Eds.), *Taal 2007: The 12th World Lake Conference* (pp. 868–881). Jaipur, Rajasthan.

Kundu, R. (2016). Making sense of place in Rajarhat New Town: The village in the urban and the urban in the village. *Economic and Political Weekly, LI*(17), 93–101.

Kundu, R. (2017, August). The "invisibles" in New Town Rajarhat: The politics of place-making by new migrants and the internally displaced refugees of urban development. Paper presented at the Sixth Critical Studies Conference on Refugees, Migrants, Violence & The Transformation of Cities, Kolkata, India. Retrieved from http://www.mcrg.ac.in/6thCSC/6thCSC_Full_Papers/Ratoola .pdf.

Lokgariwar, C., & Dewani, U. (2016, June 22). Threats to East Kolkata wetlands are threats to Kolkata: Majhi Jo Nau Dubaaye… [Web log post]. SANDRP. Retrieved from https://sandrp.wordpress.com/2016/06/22/threats-to-east-kolkata-wetlands -are-threats-to-kolkatamajhi-jo-nau-dubaaye/.

Mahadevia, D. (2011). Branded and renewed? Policies, politics and processes of urban development in the reform era. *Economic and Political Weekly, 46*(31), 56–64.

Mol, A. P. J., & Spaargaren, G. (1993). Environment, modernity and the risk-society: The apocalyptic horizon of environmental reform. *International Sociology, 8*(4), 431–459.

Narain, V. (2009). Growing city, shrinking hinterland: Land acquisition, transition and conflict in peri-urban Gurgaon, India. *Environment and Urbanization, 21*(2), 501–512.

Peet, R., & Watts, M. (Eds.). (2002). *Liberation ecologies: Environment, development, social movements.* London: Routledge.

Procter, I. (1982). Some political economies of urbanization and suggestions for a research framework. *International Journal of Urban and Regional Research, 6*(1), 83–97.

Ray, S. (2012, July 17). Wetland being sold in pieces. *The Times of India.* Retrieved from https://timesofindia.indiatimes.com/city/kolkata/Wetland-being-sold-in -pieces/articleshow/15011182.cms.

Robbins, P. (2012). What is political ecology. In P. Robbins (Ed.), *Political ecology a critical introduction* (pp. 14–16). Chichester: J. Wiley & Sons.

Sassen, S. (2005). The global city: Introducing a concept. *Brown Journal of World Affairs, XI*(2), 27–43.

Sengupta, U. (2013). Inclusive development? A state-led land development model in New Town, Kolkata. *Environment and Planning C: Government and Policy, 31*(2), 357–376.

Shatkin, G. (2016). The real estate turn in policy and planning: Land monetisation and the political economy of peri-urbanisation in Asia. *Cities, 53*, 141–149.

Sovan in Charge of Wetlands. (2017, February 21). *The Telegraph*. Retrieved from https://www.telegraphindia.com/states/west-bengal/sovan-in-charge-of-wetlands/cid/1397934.

Stott, P., & Sullivan, S. (2000). *Political ecology: Science, myth and power* (pp. 4–5). London: Arnold.

The East Kolkata Wetlands (Conservation and Management) Act. (2006). *West Bengal act VII of 2006*. (No. 404-L of 2006 dated 31st March 2006).

The East Kolkata Wetlands (Conservation and Management) (Amendment) Act. (2017). *West Bengal act V of 2017*. (No. 304-L of 2017 dated 17th March 2017).

Walker, P. A. (2006). Political ecology: Where is the policy? *Progress in Human Geography*, *30*(3), 382–395. https://doi.org/10.1191/0309132506ph613pr.

Watts, M. J. (2000). Political ecology. In E. Sheppard & T. J. Barnes (Eds.), *A companion to economic geography* (1st ed., pp. 257–274). Oxford: Blackwell Publishing.

6 Land Ownership and Ecological Knowledge Production from a Gender and Power Dynamics Perspective in a Village in Nagaland, N.E. India

A. Wati Walling

6.1 Introduction

The romanticization of tribal society as egalitarian is analyzed at the back-drop of rather contrary practices such as customary land ownership, gender and power equations, especially in socio-economic and resource management activities. An analysis is made based on field data of a constitutionally empowered village and its council or *Putu*,[1] as it is called in Yimti[2] village. Its role in governance, amendment and execution of customary practices (in regard to land ownership) are examined across all clans (nine) in Yimti village in Nagaland.

A unique dimension of this power dynamic is noted in the customary cultivation practices, where the powers that be prevail over the conventional wisdom of cultivation and *Jhum*[3] cycle. Land alienation and economic isolation were examined closely, wherein one of its root causes was located in a 'lending' or 'mortgaging' – customary practice. An overview of the origin of the land settlement and land ownership in Yimti is studied with an aim to understand the past and present pattern of settlements and its change. Land tenure, traditional land use and land holding patterns, as well as the overarching customary practices are analysed, and their intricacies in the creation of 'landlessness' and multi-layered division in Yimti across socio-political, economic spheres are outlined.

A linkage between land alienation with gender and economic isolation of the Yimti women folk is located in *Patio*[4] (Betel leaf) trade. Subjugation of the villagers happening at the level of knowledge production is observed, wherein certain forms of Traditional Ecological Knowledge (TEK) are privileged and become dominant while others are left to oblivion. It is in line with a political ecology perspective which

DOI: 10.4324/9780367486433-6

gives credence to 'multiple voices'. This paper shows the need for collaboration and synergy of knowledge systems rather than an appreciation of just one form of ecological knowledge. A political ecology perspective allows us to see how a politicized universe, or in other words, unequal power relations, change not only the politics, ecology and socio-economic fabric of a given society, but also affect broader research fields. Thus, the present inquiry is an attempt to locate linkages of land ownership, ecological knowledge production from the political ecology perspective of gender and power dynamics. This ethnographic study is situated in a 158-household village called Yimti in Nagaland in the North Eastern Region (NER) of India. Against this backdrop, the objective of this paper is to analyze how a tribal society in the contemporary North East Region of India faces concerns over land and natural resource use. Who owns, controls or influences land and its resources in the political economy of a village? What are the linkages between different levels of resource knowledge and ownership, power and governance?

6.2 Theoretical Lens, Field, Methods and Methodologies

As Forsyth (2008)[5] points out, 'social values and environmental knowledge are co-produced', and this paper centres on overarching customary land-ownership practices and the nuances therein, which adversely impact social life and the relationship with nature. Blaikie opined that:

> Political Ecology (PE) is not just a question of a comprehensive and intellectual satisfaction method for studying soil erosion, but the approach here is in direct conflict with both the dominant conventional wisdom about soil erosion ... and with the institutions charged to deal with it.[6]

In the context of this study, I argue that the 'dominant conventional wisdom' is in line with the dominating policies of the state and the power influences that be, which are in most cases in direct conflict with the local and indigenous wisdom of the people. However, in the case of Yimti, the institutions endorsing and carrying out the policies are both from within as well as outside of the village.

According to Blaikie, this paradigm shift of being sceptic towards the accepted and dominant conventional wisdom is decidedly political. Meanwhile, Peet suggests that 'empiricism itself was political, and researchers should not accept orthodox explanations of problems from physical science or expert agencies uncritically'.[7] In other words, PE can make space and time for social and environmental justice and local initiatives' (Blaikie,

2008).[8] Hence, Blaikie's approach to PE represents an integration of environmental knowledge and social justice.

This reconstruction of an environmental explanation and interventions in favour of vulnerable people can be seen as a major shift in the reframing of environmental sociology. Empirical challenges to environmental narratives have mainly come from studies of marginalized people who are delegitimized under environmental narratives, such as shifting cultivation and hill farmers, etc. 'This kind of political ecology also endorses a normative agenda to research that allows socially vulnerable people to participate in shaping future knowledge generation' (Forsyth 2008).[9] I employ political ecology (PE) as an approach to look at a micro-level interaction within the nucleus of a village. In analytical terms, political ecology 'provides tools for thinking about the conflicts and struggles engendered by the forms of access to and control over resources'.[10] (On the other hand, it is necessarily complex, as it embraces issues ranging from gender relations, land rights and local systems of governance to the socio-cultural values of environmental resources.)

All ecological issues perhaps bear a social dimension, which is an equally – if not more – important aspect of sustainability one could think about. Hence, to talk of a pristine ecology equally demands for a socially just alternative. One of the most distinctive themes in the writings of Piers Blaikie (on PE) over the years is a strong political imperative and desire to correct social injustices. Blaikie (1985)[11] asserted, '(This) is not a neutral book. It takes sides and argues a position because soil erosion is a political-economic issue, and even a position of so-called neutrality rests upon partisan assumptions'. Moreover, Blaikie tried to show that changing these values, or diversifying the social framings of environmental analysis, may result in more socially just environmental knowledge and policy. Take, for instance, 'the most talked about ecological issue association with tribal community is undoubtedly *Jhum* cultivation' (Forsyth, 2008).[12] However, the knowledge produced in the process of clearing forest, preparing and tilling ground, sowing, weeding, harvest and finally consumption is perhaps not seriously looked into. The village community life centres on 'cultivation'. And to talk of an alternative to *Jhum* as some scholars would suggest would mean a complete whitewashing of a community itself. At the same time, it also serves an equally important caution to the alternate mode of cultivation to be socially and environmentally sustainable.

Yimti is a typical Naga village situated in the Mokokchung district of Nagaland in North East India, which is an area without much significance. Economically, Yimti has a lack of profitable economic activity and hence a lack of development, politically (a case of delimitation and no concrete political party supporter), socially (unusual division of church within the same denomination for 23 years in the past) and geographically (deteriorated road to the nearest town).

Some of the methodological tools for primary data collection were participant observation, in-depth interviews, focus group interviews, government records, personal diaries and the analysis of a few elderly respondents' life history. Ethnographic research demands strong participant observation as a methodological tool, and my research work is an outcome of almost a year's stay in Yimti village with as many as four phases of visits. In sum, this analysis hinges on the historical evidence of Yimti village on one hand, and on the other the quest for the nuances behind 'landlessness' leads us to a much broader understanding of knowledge production at varied levels, especially deprivation of local knowledge and marginalization of another group within a village.

This reframing has arisen, in part, because many empirical challenges to environmental narratives have come from studies of marginalized people who are delegitimized under environmental narratives (such as shifting cultivation and hill farmers). In addition, many political ecologists have tried to empower socially vulnerable groups by carefully participatory research or by building political arenas where they can speak.[13] Consequently, rethinking knowledge or social order may therefore allow the creative or positive reconstruction of both environmental understanding and politics, in favour of vulnerable people.

6.3 Land, Customary Practices and State Policies

The value of land tends to be priceless, more so when it comes to an agrarian community whose identity, status and livelihood centres on that land. The land-owning practice of the Naga tribe is unusual, distinct and intricately linked with customs and traditions which are remarkably diverse in nature. One reason for this diversity could because of the distinctly assorted linguistic tribal groups. There are 16 major tribes with an entirely different dialect and scores of other minor tribes in Nagaland. It can be noted that each tribe has its own distinct customs and traditions in the form of different festivals, ways of cultivation, land ownership, settlement, housing patterns and governance, etc. It was only in the mid-twentieth century that a good number of Naga tribal groups started following Christianity, which brought about social awareness and education. Prior to Christianity, by and large all the tribal people lived in isolation and with much hostility towards each group. Today, the tribals are the custodians of 88.32% of the total landmass of Nagaland. Customary land-ownership practices impact identity formation across different categories within households, clans, gender, religion and other social institutions. One of the existing customary grain-lending practices is perceived as an agent which has brought about land alienation and thus a massive out-migration in Yimti village in 1964 and division within a church of the same Protestant denomination from 1993 until 2016. Credibly, a sense of identity formation between the land owners, custom practitioners and the landless tribals is seemingly created. By and large, the tribal communities in Nagaland retain their uniqueness through an ongoing oral tradition. Therefore, with its rather elusive written history, a description of ethnography helps to enquire upon individuals and institutions of the past and the present Yimti village.

Nagaland has a unique type of land ownership, where the land is owned by individuals, clans and the village community. On the other hand, there is the presence of customary law in the state, which can even bypass amendments and rules laid down by the Centre Government of India. Today, villagers are the undisputed owners of almost the entire land in the rural areas. Except for some lands which were acquired by the government for specific purposes of common interest either through payment, land compensation or through donation from villagers. Some transactions of land did occur in the past between the state and individuals which require survey and an updating of land records, even in Yimti. In the meantime, local customs and traditions are strongly opposed to the government's land records system as they perceive this exercise to be a threat to their long-standing concepts of private property rights. Any attempt to survey and map boundaries and exact land is looked upon as an evil design of the government to alienate them from their land or a prelude to levying land taxes. Hence, no co-operation is extended by the villagers in most cases while a surveying attempt was made by government officials. This is perhaps so because of the traditional way of living in isolation and non-interference between different tribals groups.

6.4 Land-Owning Practices

Nagaland state has a distinct land-ownership pattern in owning 88.32% of the total landmass and a meagre 11.68% belongs to the Nagaland government. Land-owning patterns among Naga Tribes can be broadly grouped into: those with chieftainship and those without chieftains. In a chieftain-based village, the chief plays a pivotal role in the affairs of the community. No individual or group is recognized as owners except for the chief himself. There are, however, differences within institutions governed by chiefs.

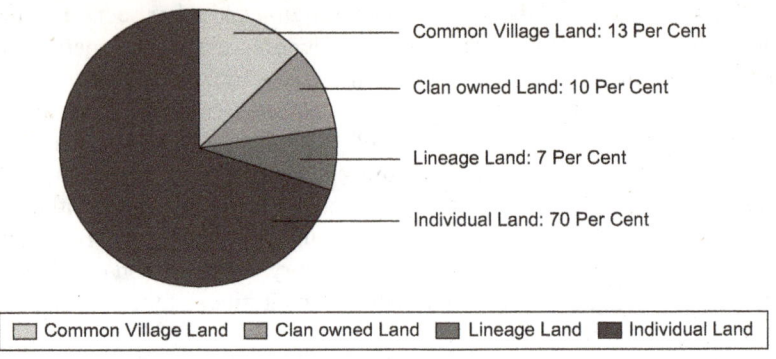

Common Village Land: 13 Per Cent

Clan owned Land: 10 Per Cent

Lineage Land: 7 Per Cent

Individual Land: 70 Per Cent

☐ Common Village Land ☐ Clan owned Land ▨ Lineage Land ■ Individual Land

Figure 6.1 Pie chart indicating distribution of land ownership in Yimti Village.

A village without chieftains generally has the following classification of land ownership. Yimti village falls the under non-chieftainship category.

6.5 Clan and Land-Ownership Relations

Out of the 11 clans, 9 clans are taken for this analysis as the remaining 2^{14} are in Yimti village only for a specific time and purpose thus, does not possess any land or have any extended families. The two clans Chari (one household) and Longkumer (two households) is therefore, not a part of this analysis. There are only three clans in Yimti who possess very large plots of land. They are Aier, Ozukum and Tzudi. Tzudi Clan is not only more in population but also owns a good portion of land. However, Tzudi clan also has the highest number of households who are landless. Out of the nine clans there are only two clans, namely Pongen and Lemtor who do not have any households without land. Clan is where a person is born into and cannot be changed. Hence, name of the clan become part of one's own identity by means of having it as a part of one's own name. 'In our society, clan is one thing a person takes pride and so women folk chose to retain their clan name (in the form of surname/last name) even after marriage'.[15] Studies show that as much as 13% of the population of 158 households are either landless or own an insignificant amount of residential land. Tracing back to the past, on one hand every household possessed land, and on the other hand the state's intervention of prohibiting non-indigenous residents to own land. This directs us to question whether such emergence of landlessness is from within the community (Table 6.1).

Table 6.1 Distribution of land ownership according to clan

Sl. No	Clan name	Landless/ very small[a]	Moderate[b]	Large[c]	Very large[d]	No. of households (HH)
1	Lemtor	–	5	1	–	6
2	Sempo	2	5	–	–	7
3	Longchar	3	3	2	–	8
4	Pongen	–	9	1	–	10
5	Aier	4	6	–	1	11
6	Walling	1	9	1	–	11
7	Jamir	1	18	1	–	20
8	Ozukum	4	23	8	2	37
9	Tzudi	6	32	6	4	48
Total		21	110	20	7	158

[a] Land holding up to 3,600 sq ft or less; usually residential area and no agricultural land.
[b] Land holding up to 1 bigah = 14,400 sq ft (Bigah calculation according to the scale of the respondents). Mostly agricultural land.
[c] Land holding up to 1 pura or 57,600 sq ft.
[d] Land holding beyond 1 pura or 57,600 sq ft.

6.6 Power Dynamics and Customary Cultivation Practices

The authority over land in Yimti village seems to rest exclusively on the individuals since a major portion of the land is privately owned. However, on the ground level, the village council or the customary law guardians directs the entire scenario in an interesting fashion. Traditionally, Yimti villagers are by and large dependent on agriculture for their livelihood. Farming activity in Yimti is exclusively slash-and-burn or *Jhuming* cultivation, although some irrigated rice terraces are found in scattered parts. A particular *tract*[16] is selected for this purpose and is cultivated over a two-year period, which then lies fallow for several years, allowing new jungle to grow up. The forest, thus earmarked for cultivation, is cleared. Both men and women do the sowing. Interplay of the influential and the *Putu* (village council) is found to exist in choosing a particular tract for the two-year cultivation. This is crucial because even though it's private-owned land, a household or two cannot decide on a tract since it would mean clearing and renovating the entire footpath to the tract. Hence, traditionally it is on *Putu* to decide the tract and carry on the footpath construction process every third year. Manen[17] has to say that

> this decision of the Putu to locate a tract is crucial because it is almost the entire villagers whose labour would be rendered for at least a couple of days to clear the particular route to reach the identified tract.

Shiluti[18] too asserted that, 'unlike olden days, where cultivable tract was identified predictably and systematically, lately, there has been influence of the *Putu* by certain individuals who have ulterior motives'. Clearly, dilution of customary practices of tract identification through traditional wisdom is overlooked and the land does suffer loss due to unreasonable cycles of keeping it fallow. There is also bourgeoning unpredictability in terms of harvest due to the same reason of land misuse.

Hence, the power dynamics are played out in the way the village council (*Putu*) decides which portion of the land should be cultivated as per the *Jhum* cycle. Secondly, *Jhum* cultivation involves great risk, especially while burning the cleared forest. Therefore, it has to be a group activity and not just a few households or an individual decision to cultivate where or which one of his lands is to be cultivated. It is so, as it calls for unforeseen dangers of burning the entire forest or even spreading fire in the village (which is a great offence according to the customary law). Thirdly, in order to cultivate a specific plot of land, there has to be a proper footpath. This again calls for large-scale labour which can only be mobilized by the *Putu* and not just few individuals or households.

It has been found that just a handful of households own more than half of the total land in Yimti village. These land owners are invariably the ones who grabbed land from the grain borrowers. However, when it comes to political affiliation they are all opportunist with no group or clans remaining faithful to a political party. This is perhaps the implications of constitutional delimitations which do not recognize Yimti village to be part of the 'range'[19] that she is in.

6.7   Betel Leaf (*Patio*) Trade: Gender Isolation and Power Equation

This economic activity of the rich and the landed is multifaceted. Betal leaf or the locally know *Patio* found in and around Yimti villages is said to be delicate and tasty. Patio trade occupies a special place in revenue generation and also draws insights on unsuspecting gender and power equations. Yimti village has a tradition of *Patio* trade, where business people from nearby towns such as Tuli, Amguri (Assam) and Mokokchung come to the village to buy *Patio*. Yimti village produces high-quality *Patio*. It stands highest next only to Longjang village across the Mokokchung district of Nagaland in terms of quality and export. It is also noted that *Patio* business could have been one of the reasons behind the Yimti villager's good contacts with Ahom (Assam) kingdom – a trade tracing back to more than 100 years. However, the timber (log) business in the 1990s had an adverse impact, when trees were felled for the timber business since *Patio* is a climber species and needs prominent trees to grow and such trees were felled for timber. *Patio* leaf per bundle costs up to Rs. 50 (negotiable) for normal ones and Rs. 60 for the best quality. A bundle has 10 khati and a khati has 17–18 leaves (which are supposed to be 20, according to Meren).

There were about five groups in Yimti village who are involved in *Patio* trade. Each group consists of 7–8 Tekidars. Tekidars also go around collecting the leaves in their respective *Kiyong* (*Khel*) where they struck deals such as the rates and commissions, etc. It is found that sellers (villagers) do not have loyalty to any Tekidar but most of the villagers look out for a better bargain. It was noted that Tekidars used to get commission of 5–10%, and Rs. 50 extra as overall commission money. Meren further opined that

> Given a harvest from a healthy vine, approximately Rs. 600–Rs. 700 could be earned from a tree. Therefore, on an average, a large betel leaf grower earns up to Rs. 35,000/- solely from this activity which spans for 4 months.

According to an elderly person[20] 'there is a need to regulate (perhaps by the Village Council) these *Patio* business groups and Tekidars to just 2–3, and also impose penalty on the misappropriation of the leaves by Tekidars as well as the villagers'.

This rather lucrative venture in the context of Yimti village is another example of economic isolation of the land owners from the common people. Here, the land owners who grow betel vines make use of the cheap labour of the landless villagers for collecting *Patio*. It is also seen that most of the Tekidars are either the sons of large landowners or their close relatives. Although women folk are equally involved in the labour process of picking betel leaves and cleansing, there was no record of women Tekidars in the village thus far. This economic alienation needs to be understood in the broader political ecological perspective of control over resource access. On one hand, the domination over the vulnerable landless subjects, on the other hand, namely the women folk and the landless villagers. The whole idea of regulating the number of Tekidars, which is likely to happen in the near future, is another instance of how the nexus of the influential monopolizes the trade and resource control.

Throughout the tribal village studies in India and abroad, a component of gender can be perceived conspicuously. It was recorded that men participate actively in politics, and women participate passively (Ishwaran 1968: 34, quoted in Jodhka).[21] In yet another instance, women were thought to be incapable of understanding what went on outside the domestic wall (Srinivas 1976).[22] Analysis of the bilingual nature of man and mostly native language spoken by women, etc., were attributed to the different social roles between men and women. In the case of Yimti village, it was noted that the '*Zunga*'[23] system played a great role in bridging the gender gap and gender-related issues such as insulating women from socio-political activities, etc. *Zunga* groups are composed of both genders, who work in tandem with other *Zunga* groups in the village. In this process, the *Zunga* system ensures women's participation up to an extent and also finds women involvement across different *Zunga* groups. Nevertheless, within the superficial super-structure of the egalitarian village community created and regulated by mechanism such as *Zunga* system, the deeper economic and gender issues in politics are yet to be realized.

> Naga society is a patriarchal society where property is inherited by the sons, starting from the eldest son. It can be argued that 'the exclusion of women from the line of inheritance, ownership and control of resources confers on them a secondary status in the agrarian social structure and contributes to their overall subordination in society in spite of their economic contribution.
>
> (Chowdhry 2009)[24]

Clearly, the ownership and control of land signifies power equations among agricultural communities. Every society, therefore, develops its own norms of inclusion and exclusion for the determination of the nodal points of power, authority and the control of resources. Traditionally land has been the basis of political power and social status. For many, 'it [land] provides a sense of identity and rootedness within the village, and often in people's mind a land has a durability and permanence which no other asset possesses' (Selvadurai, as quoted in Agarwal, 1996).[25] In any case, 'ancestral land usually has a symbolic meaning which purchased land does not give, within some village communities, continuity of ancestral land also stands for continuity of kinship ties and citizenship' (Agarwal, 1996).[26] This phenomenon is found to be true in the case of Yimti village, where several households seem to have lost their land to debt and in the process face humiliation and a loss of identity in their own ancestral land. On the other hand, women's status is equally deprived when it comes to land ownership, governance and control of land, where women are neither allowed to inherit or own nor take part in the governance matters.

However, the church seems to come to the rescue of women folk.[27] According to Satemla,[28]

> the solidarity of women is rooted only in church where there is church services exclusively meant for the women folks (Every Thursday 8–9pm). This Thursday service[29] is attended mainly by middle aged

housewives though women of any age group could attend the same. The service is initiated and looked after completely by the women pastor of the church. This worship service is meant for the teachings from the bible. However, the idea of getting together in exclusivity gives women a fair space to discuss and share their household matters as well as concerns in regard to the family, church and community at large.

This solidarity of the women in the place of worship (church) could be seen as a means to raise the voice of the women folk, where there is no other platform for the women to address and complement to the governance and negotiations in the village administration. All in all, the church has significantly institutionalized an authenticity to the women folk which no other institutions could provide. However, a woman leader of a church has comparative authority limited within the church and they do not have a direct role in the overall church administration.

6.8 Customary Grain-Lending Practice and Land Alienation

Until today, almost all the households of the Yimti village practice *Jhum* cultivation in one's individual land or clan-owned land. At the time of research, it was found that about 87% of the Yimti villagers owned land. However, the scenario of land ownership tipped down in the 1960s where land ownership by certain individuals rose astronomically, leaving a majority of households without land. The most exploitative system of land alienation for Tali[30] is the standard customary practice of lending grain in Yimti. In the past, there were instances of famine and epidemics in Yimti village. Along with such tragedies, there were households that would not harvest sufficiently for the whole year. During such instances, many households were left with no option but to borrow grain for their square meals.

The grain-borrowing system was usually measured by a tin container.[31] As per a customary practice, while lending grain, the lender would seek for an agreement of mortgaging the land from the borrower with a dateline. According to a customary practice, on completion of a specific time (normally a year from the time of borrowing), the borrower of a tin of grain (12 kgs), has to pay back half a tin extra as interest (that is, $12 + 6 = 18$ kgs) in return. The deal gets settled if the debtor pays back the right measure of grain on time. However, if the debtor cannot settle the complete measure on time, then the amount of grain becomes double to that of the current debt, i.e., by the end of the second year, the person has to return three tins of grain. ($18 + 18 = 36$ kgs, which is three times the amount borrowed two years ago). On the occasion of failing to repay this measure, the debtor is forced to relinquish his property without a further chance to redeem his mortgaged property (mostly land). Repayment of debt was always difficult as the villagers solely depend on just one harvest through the calendar year. This system of exploitation has been the weapon of few powerful rich in the name of customary practices to capture and accumulate numerous plots and cultivable land for themselves. Till today agriculture is the major, or rather the only, source of sustenance for many villagers of Yimti, without surplus for most of the households. It is an accepted phenomenon that almost every year some households in the village would harvest insufficiently due to the diminishing returns from agriculture and virtually no produce from the wild lately. This is one reason for how the loss of land in Yimti village began. It may be noted that migration to urban areas since the attainment of statehood of Nagaland in 1963 for employment is also coupled with the exploitative nature of customary practices, such as the grain-lending practice which often leads to a loss of landed property.

6.9 Competing Knowledge Production for Policy Appropriation

The modernization crisis has reached such a saturation point today that its critical evaluation has become vital and very crucial. This necessarily leads to the fact that, to ignore people's knowledge by and large ensures failure in development. And for the experience of Yimti village it can be seen that a policy totally devoid of villager's knowledge is rather seen as foreign and likely to fail. Perhaps people's knowledge is essential for development and hence it is often suggested that it must be gathered and documented in a coherent and systematic fashion. As more studies of Traditional Knowledge (TK) have taken place its value and relevance have begun to seem self-obvious. Some of the reasons which uneven development has brought about are much too conspicuous today – the persistence of poverty amid increasing affluence and the increase of employment being one of them. One of the reasons is to be attributed to the uneven distribution of a knowledge base.

Unlike any developed countries whose knowledge on environmental degradation is on protection, conservation and sustainability, the Third World is directly related to livelihood issues. Above all these, according to Kaldor,[32] 'there is a loss of hope – an erosion of the myth that development can create a just and humane society'. And this erosion has on the other hand permitted the popularity and self-assurance of a non-Western social, cultural and political movement.

Appadurai (in Frederique and Marglin, 1990: 96) asserted that, in many of the traditional agrarian societies and some non-agrarian societies too, it is difficult to distinguish technical knowledge very clearly from knowledge which is tied to larger normative and social ends. It is imperative, therefore, to know the local community's significant differences in land ownership, ecological knowledge production and the gender and power dynamics within. Traditional Ecological Knowledge (TEK) in a sense means the local ecological knowledge and skills which people construct on their own, over a certain period of time. TEK is unique to a given culture, which is by and large contrasting to international knowledge systems which are generated by universities, research institutions and private firms. Such knowledge is passed down from generation to generation in many societies by word of mouth. Yimti village is one such community whose knowledge has been limited to traditional verbal means of communication. TEK, according to them, has value not only for the culture in which it evolves, but also for scientists, policy makers and planners striving to genuinely improve conditions in rural localities. It is the basis for local-level decision makers in agriculture, ecological conservation, education, natural resource management and a host of other activities, mainly in rural communities. It is also obvious in an agriculture-dependent society since the people are involved in activities of that kind, they also possess and use their knowledge, which is instinctive in the absence of other external knowledge.

According to Warren, 'Indigenous knowledge (IK) is an important natural resource that can facilitate the development process in cost effective, participatory, and sustainable ways' (as quoted in Flury & Geiser et al., 2002).[33] IK also values ritual and spiritual dimensions which, in more cases then not, preserve nature. Awareness of indigenous knowledge (IK) comes as a wakeup call for its potentiality towards sustainable development and poverty elevation. However, there has been a lack of appreciation and serious consideration as IK as such is not supported by written documents, rules and regulations and technological infrastructure. The recent danger of disappearing IK is not just the influence of global processes and rapid change, but also because the efforts and facilities are needed to be documented, evaluated, validated, protected and disseminated. Nevertheless, not all IK offers a sustainable solution to today's pressing and complex problems. However, a deeper appreciation of nature–society dynamic of people living in the remote regions such as Yimti village may offer perspectives to alleviate poverty and marginalization for those whose alternative form of sustained growth is bleak.

6.10 Bridging the Gap: A Way Forward

Be it competing ecological knowledge, gender and power dynamics emerging out of land use, it is inadvertently an argument surrounding traditional and Western/scientific knowledge. Does traditional knowledge have significance towards development, equity, justice, basic needs and further economic growth to sustainable development by means of absorbing it for policy making?

It is observed in the earlier experiences that traditional or the so called traditional knowledge and institutions have been considered as obstacles to development. In fact, the customary cultivation practices and power play of the power that be has been analysed in this paper. However, distinction between customary practices such as the grain borrowing practices, selection of tract for cultivation, etc. and traditional knowledge of understanding the land and surrounding ecology needs to be evaluated and appreciated. Arguably, the beginning of formal colonial education itself marked indigenous methods as native, simple, and primitive and colonial methods as European, modernizing and civilized.

(Radcliffe & Edward, B., et al. quoted in Warren, 1995)[34]

However, in the words of Agrawal (1994),[35]

the attempt to create distinction in terms of indigenous and western is potentially ridiculous. It makes much more sense to talk about multiple domains and types of knowledge, with differing logic and epistemologies. Knowledge is necessarily anchored in institutional origins and moorings where protection, systematization and dissemination of knowledge are useful to different people in different ways.

It is only when we move away from the dichotomy of traditional and Western knowledge that a productive dialogue can be ensured, which will focus on safeguarding the interests of those who are disadvantaged. If science is the ultimate arbiter of knowledge, then there seems little point in advocating for the distinction between scientific and traditional knowledge. However, the way out and the middle path would perhaps be a participatory synergistic approach for policy research, formulation and implementation.

In the midst of an increasing gap between the vision and reality of participatory research for and of the people, there is a need to find ways to narrow down conflicts between "sustainable livelihoods and requirements for sustainable resource management" (Baumgartner et al., 2002).[36] Participatory research has the potential to strength cultural identity and contribute to a community's ability to cope with social change. It also means narrowing down the diverging interest between the powers that be and the citizens. One such technique is the research feedback technique.

It is a method based on an entry-level interaction with the people, with their feedback or opinions on a particular case in point. Research feedback as a method has the potential to promote endogenous learning within the boundaries of existing indigenous knowledge and to identify 'tacit source of knowledge'.[37]

Therefore, in the context of this paper, Ecological knowledge production through the powerful *Putu* is observed. Marginalization and the alienation of the landless and the female gender in socio-economic activity such *Patio* trade are some instances. It is true that the government state's law and policies are rather illusive and surreptitious in nature and at best could be attributed as 'negligent' due to the conflict-ridden history and the political undercurrent of the state. A theoretical paradigm like political ecology allows us to enquire upon a much-politicized world from the point of view of the vulnerable section of the society. This paper therefore suggests that people became vulnerable to poverty and landlessness in Yimti for reasons such as politicization of ecological knowledge by controlling customary cultivation practices, customary grain-lending practices which sweepingly favoured the landed villagers up to the extent of forfeiting their only possession – land. The socio-economic alienation of the women in one of the most lucrative *Patio* businesses in the village is also recorded. Indeed, there is a pressing need for the state to address exploitative customary practices, such as the one which tends to subjugate certain section of society, namely the women and the landless. It is perhaps pertinent that the state ensures that its policies promote equitable and sustainable development by taking into account the inhabitants' knowledge narratives, engaging them in participatory research. There are emerging studies which support the claim of dissatisfaction with the modernizers' perspective on tribal society, and in order to understand the current crisis (both ecological and societal), and to discover reasonable and coherent alternatives, it is necessary to examine in detail the intellectual and cultural roots of policies and the knowledge embedded therein.

Notes

1 *Putu Menden* or *Putu* is the name for the governing official and member as well as the office of the village council.
2 Yimti village is a pseudonym for a village in Mokolchung district in Nagaland which is predominantly inhabited by the Ao-Naga tribe. Ao tribe is 1 among 16 major tribes in the state of Nagaland.
3 Form of cultivation popularly found in the entire North East region of India and exclusively found in the research field area – Yimti village.
4 Or Paan leaf. It is a climber plant species and its leaves are used extensively among the North-East tribals (and beyond) as a mouth refresher and traditional paan.

5 Forsyth, Tim. 2008. *Political Ecology and the Epistemology of Social Justice.* Science Direct, Elsevier. Geoforum: 39 (2008), pp. 756–764, Elsevier, p. 756.

6 Blaikie 1983; 29 as quoted in Forsyth, Tim. 2008. *Political Ecology and the Epistemology of Social Justice.* Science Direct, Elsevier. Geoforum: 39 (2008), pp. 756–764, Elsevier, pp. 756–757.

7 Ibid., p. 757.

8 Blaikie, Piers, 2008. *Epilogue: Towards a Future for Political Ecology that Works*, Science Direct, Elsevier. Geoforum: 39, pp. 765–772.

9 Forsyth, op. cit., p. 757.

10 Peluso and Watts 2001: 25 in Jewitt, Sarah. 2008. *Political Ecology of Jharkhand Conflict.* Asia Pacific Viewpoint, Vol. 49, No. 1, April 2008. p. 69.

11 Blaikie, Piers. 1985. *The Political Economy of Soil Erosion in Developing Countries.* London. Routledge, p. 1

12 Forsyth, op. cit., p. 757.

13 Escobar, 1996. In Forsyth, Tim. 2008. *Political Ecology and the Epistemology of Social Justice.* Science Direct, Elsevier. Geoforum 39 (2008), pp. 756–764, Elsevier, pp. 756–757.

14 They are: one Chari household and two Longkumer households, who came to the village on government posting.

15 Wapangla is a 48-year-old woman, interview on 28 March 2019 at Yimti village.

16 The entire village's cultivable land is divided into several tracts. One tract would normally accommodate a large portion of the villagers, if not all of them, to cultivate for a consecutive period of two years.

17 -year-old man, interviewed on 21 March 2019 at Yimti village.

18 -year-old man, interviewed on 21 March 2019 at Yimti village.

19 Mokokchung district is divided into six major ranges which finds common identity, especially during the time of election.

   Meren is 33-year-old young man interviewed on 28 March 2019 at Yimti village. He further hinted that Patio trade should be handed over to the *Putu* so that no discrepancies and dishonesty of this sort happens in the village.

   Tekidars are vendors, who collect betel leaf from the villagers (betel leaf growers) and hand it to buyers. These buyers from Yimti village then sell to other businessmen in their respective towns.

20 Sungkum is a 77-year-old man interviewed on 2 April 2019 at Yimti village.

21 Jodhka, Surinder. S. *From "Book-view" to "Field-View": Social Anthropological Constructions of the India Village.* QEH Working Paper Series. (unpublished), p. 34

22 Srinivas, M.N. 1976. *The Remembered village.* Delhi: Oxford University Press, pp. 140–141

23 'Zunga' (in Ao-dialect) refers to age grouping system found in Yimti village and beyond. This grouping happens across gender and is stringently hierarchical in nature.

24 Chowdhary, Prem (eds). 2009. *Gender Discrimination in Land Ownership*, New Delhi: Sage Publications, pp. 75–78.

25 Agarwal, Bina. 1994. *Field of One's Own: Gender and Land Rights in South Asia.* Cambridge: Cambridge University Press, pp. 17–19.

26 Ibid., pp. 17–19.

27 This women solidarity has also been impacted due to the emergence of landlessness in the village. This is perhaps one of the adverse effects of division which has a direct impact on women.

28 Satemla is a 55-year-old woman, interviewed on 3 April 2019 at Yimti village.
29 Service in Christian context is a time of devotion inside the church building. This 'service' or 'worship' as it is called is primarily for teaching from the Bible and singing hymns, etc.
30 Tali is a 60-year-old man interviewed on 1 April 2019 at Yimti village.
31 A measure of a tin of grain = 12 kgs approximately. There are other lesser measures but this is considered standard for accuracy and convenience.
32 As quoted in Marglin and Marglin 1990, p. 31.
33 Flury, Manuel and Geiser, Urs (eds). 2002. *Local Environmental Management in a North-South Perspective: Issues of Participation and Knowledge Management.* Amsterdam: IOS Press, p. 162.
34 Warren, D. Michael, et al. 1995. *The Cultural Dimension of Development, Indigenous Knowledge System.* London: Intermediate Technology, pp. 35–37.
35 Agarwal, Bina, 1994. *Field of One's Own: Gender and Land Rights in South Asia.* Cambridge: Cambridge University Press, p. 31.
36 Baumgartner R., Aurora, G.S., Karanth, G.K. and Ramaswamy, V. 2002. *Researchers in Dialogue with Local Knowledge Systems – Reflections on Mutual Learning and Empowerment,* pp. 255–274, in V. Flury, Manuel. V. and Urs Geiser (eds) 2002. *Local Environmental Management in a North-South Perspective: Issues of Participation and Knowledge Management.* Zurich: IOS Press.
37 Ibid., p. 54.

Bibliography

Agrawal, Arun. 2002. *Indigenous Knowledge and the Politics of Classification.* Oxford: Blackwell Publishers.

Agrawal, Arun and Sivaramakrishnan, K. 2001. *Social Nature: Resources, Representations and Rule in India.* New Delhi: Oxford University Press.

Agarwal, Bina. 1994. *Field of One's Own: Gender and Land Rights in South Asia.* Cambridge: Cambridge University Press, p. 31.

Ao, Tajen. 1980. *Ao Naga Customary Laws.* Mokokchung: Aowati Imchen.

Apffel Marglin, Frédérique and Marglin, Stephen A. 1990. *Dominating Knowledge: Development Culture and Resistance.* Oxford: Clarendon press.

Baviskar, Amita (Eds.). 2008. *Contested Grounds: Essays on Nature, Culture and Power.* New Delhi: Oxford University Press.

Brara, Rita. 2006. *Shifting Landscapes: The Making and Remaking of Village Commons in India.* New Delhi: Oxford University Press.

Blaikie 1983; 29 as quoted in Forsyth, Tim. 2008. Political Ecology and the Epistemology of Social Justice. *Geoforum*, 39, 756–764.

———. 1985. *The Political Economy of Soil Erosion in Developing Countries.* London: Routledge, p. 1.

———. 2008. Epilogue: Towards a Future for Political Ecology that Works. *Geoforum,* 39, 765–772.

Cederlof, Gunnel. 2008. *Landscapes and the Law: Environment Politics, Regional Histories, and Contests Over Nature.* New Delhi: Permanent Black.

Cederlof, Gunnel, and Sivaramakrishnan, K. (Eds.). 2005. *Ecological Nationalism: Nature, Livelihood, and Identities in South Asia.* New Delhi: Permanent Black.

Eden, Sally. 1998. Environmental Issues: Knowledge, Uncertainty and the Environment. *Progress in Human Geography*, 22(3), 425–432.

Escobar. 1996. In Forsyth, Tim. 2008. Political ecology and the epistemology of social justice. *Geoforum*, 39, 756–764.

Flury, Manuel and Geiser, Urs (Eds.). 2002. *Local Environmental Management in a North-South Perspective: Issues of Participation and Knowledge Management.* Amsterdam: IOS Press.

Forsyth, Tim. 2008. Political Ecology and the Epistemology of Social Justice. *Geoforum*, 39, 756–764.

Foucault, Michel. 1982. The Subject and Power. *Critical Inquiry*, 8(4) Summer, 777–795.

Fox, Jefferson M. 2000. *How Blaming 'Slash and Burn' Farmers is Deforesting Mainland Southeast Asia. Asia Pacific Issues. East-West Center.* Analysis from the East West Center No. 47: 1–8

Gadgil, Madav. 1983. Forest Management and Forest Policy in India: A Critical Review. *Social Action*, 23(2) April–June, 127–155.

Gadgil, Madhav and Guha, Ramachandra. 1992. *This Fissured Land: An Ecological History of India.* New Delhi: Oxford University Press.

Geertz, Clifford. 1973. Thick Description: Toward an Interpretive Theory of Culture. In *The Interpretation of Cultures: Selected Essays.* New York: Basic Books.

Government of Nagaland. 2007. *Statistical Hand Book of Nagaland 2007.* Kohima: Directorate of Economics and Statistics.

Government of Nagaland. 2008. *Compilation of Acts, Rules, Notifications, Memoranda and other instruments of Government for Land Revenue Administration.* Kohima: Land Revenue Department, Government of Nagaland India.

Guha, Ramachandra. 1990. *The Unquiet Woods: Ecological Change and Peasant Resistance in the Indian Himalaya.* Berkeley, CA: University of California Press.

Guha, Ramachandra (Eds.). 1994. *Social Ecology.* New Delhi: Oxford University Press.

Guha, Sumit. 2003. *The Politics of Identity and Enumeration in India c. 1600–1990.* Society for Comparative Study of Society and History. 0010-4175/ XX/148-167.

Jodhka, Surinder S. *From "Book-View" to "Field-View": Social Anthropological Constructions of the India Village.* QEH Working Paper Series. (unpublished). p. 34.

Linkenbach, Antje. 2007. *Forest Futures: Global Representations and Ground Realities in the Himalayas.* New Delhi: Permanent Black.

Marglin, Frederique A. and Marglin, Stephen A. 1990. *Dominating Knowledge: Development Culture and Resistance.* Oxford: Clarendon Press, pp. 36–78.

Mosse, David. 2003. *Rule of Water: Statecraft, Ecology and Collective Action in South India.* New Delhi: Oxford University Press.

Peet, Watts. 1999. *Liberation Ecologies.* Berkeley, CA: University of California Berkeley.

Peluso, Nancy Lee. 1992. *Rich Forests, Poor People: Resource Control and Resistance in Java*. Berkeley, CA: University of California Press.

Peluso, Nancy Lee and Peet, Watts. 2001. 25 In Jewitt, Sarah. 2008. Political Ecology of Jharkhand Conflict. *Asia Pacific Viewpoint*, 49(1) April, 69.

Rao, Nitya. 2008. *Good Women do not Inherit Land: Politics of Land and Gender in India*. New Delhi: Orient Black Swan.

Ritzer, George. 1975. *Sociology – A Multi Paradigm Science*. Boston, MA: Allyn and Bacon Inc.

Saberwal, Vasant K. and Rangarajan, Mahesh (Eds.). 2003. *Battles Over Nature: Science and the Politics of Conservation*. New Delhi: Permanent Black.

Scott, James C. 1976. *The Moral Economy of the Peasant: Rebellion and Subsistence in Southeast Asia*. New Haven, CT: Yale University Press.

Shimray, U. A. 2007. *Ecology and Economic System: A Case of the Naga Community*. New Delhi: Regency Publications.

Srinivas, M. N. 1976. *The Remembered Village*. New Delhi: Oxford University Press, pp. 140–141.

Vandergeest, Peter and Peluso, Nancy Lee. 1995. Territorialization and State Power in Thailand. *Theory and Society*, 24(3) June, 385–426.

Vasan, Sudha. Policy, Practice and Process: Understanding Policy Implementation. *Indian Institute of Advanced Studies*. Rashtrapati Nivas, Shimla (unpublished).

Vashum, R. 2005. *Poverty of Knowledge and Its Ramifications on Indigenous Peoples: A Native Responds to the Prejudiced Writings of Outsiders*. Mainstream July 8–14. 2005: 11–14.

Warren, D. Michael, et al. 1995. *The Cultural Dimension of Development, Indigenous Knowledge System*. London: Intermediate Technology, pp. 35–37.

Xaxa, Virginius. 1999. Transformation of Tribes in India: Terms of Discourse. *Economic and Political Weekly*, 34(24) June 12–18, 1519–1524.

7 Political Ecology of Natural Resource Governance in Chhattisgarh, India

Critical Ethnographic Reflections on Vulnerable Livelihoods of the Scheduled Tribes in Bastar

Janmejaya Mishra

7.1 Introduction

Bastar division of Chhattisgarh is a highly forested and mineral-rich region in Central India. The forested landscape spreads across 39,117 sq km of geographical area and supports over 22.50 lakh population belonging to 7 administrative districts. A large majority of the populations belong to several marginalized Scheduled Tribe (ST) communities, some of whom are Particularly Vulnerable Tribal Groups (PVTGs). Human development indicators in health, education and nutrition among the tribal communities in Bastar are the lowest in the country. Subsistence of rainfed agriculture and forest-based livelihoods significantly contribute to the food security and income of the tribal households in the region. However, access to and control over natural resources such as land, forests and minerals has been a highly contested issue over the past few decades in Bastar. Degradation of forest resources resulted due to commercial exploitation of timber during the colonial period has been further intensified by various mining and development projects. Socio-economic and political marginalization of the tribal communities by moneylenders, local traders and contractors have led to wide-spread unrests among the local communities. Low level of literacy, lack of awareness among the tribals about their rights and entitlements, ineffective participation in governance process, seasonal labour migration, poor outcomes of government programmes, high level of corruption, bureaucratic indifference, absence of robust civil society interventions and Left-Wing Extremist violence are the main contemporary development challenges in Bastar.

Primarily drawing up on long-term ethnographic research in Bijapur district, this article seeks to explore the existing processes of governance of

DOI: 10.4324/9780367486433-7

mineral and forest resources using political ecology framework and critically examines the impacts of the ongoing resource conflict over natural resources and development on livelihoods of the marginalized tribal communities in Bastar, Chhattisgarh. The paper is structured into five major sections, apart from the introduction and conclusion. The following section on political ecology of natural resource governance offers a brief overview of various theoretical arguments made by the key contributors in the field of political ecology and outlines some of the essential components of political ecology of natural resource governance in the Indian context. The next section provides empirical evidence of the processes of over-exploitation of mineral and forest resources and its consequences on the tribal populations. The section on NTFP-based livelihoods analyzes the key findings of a survey on the crucial contribution of NTFPs to the income of tribal households and major issues affecting the forest-based livelihoods in Bastar. The subsequent section briefly examines the impact of Left-Wing Extremist violence and the devastation caused by Salwa Judum campaign in the lives and livelihoods of the poor tribal households alongside offering fresh insights on how effective implementation of government's flagship programmes can make a difference in the lives of the tribal populations in Bastar. Finally, the section on strengthening governance process of natural resources emphasizes the potentials of PESA Act, 1996 and FRA, 2006 in improving the governance processes of natural resources through active participation of the tribal communities in Bastar.

7.2 A Theoretical Framing of Political Ecology of Natural Resource Governance

The field of Third World political ecology originated in the early 1970s at a time when human–environmental interaction was coming under close public and scholarly scrutiny, especially in the First World (Bryant & Bailey, 1997). Blaikie and Brookfield (1987) argue that political ecology combines the concerns of ecology and a broadly defined political economy. Together this encompasses the constantly shifting dialectic between society and land-based resources, and also within classes and groups within society itself (Blaikie & Brookfield, 1987). Greenberg and Park (1994) argue that it is a synthesis of political economy, with its insistence on the need to link the distribution of power with productive activity and ecological analysis, with its broader vision of bio-environmental relationships (Greenberg & Park, 1994). Bryant (1998) suggests that political ecology examines the political dynamics surrounding material and discursive struggles over the environment in the Third World (Bryant, 1998). Guha and Martinez-Alier (1997) state that in ecological economics, the study of distributional issues constitutes a new field of study, which we call 'political ecology'. While political economy (in the classical tradition) studies economic distribution of conflicts, a new field is emerging, political ecology, which studies ecological distribution of conflicts (Guha & Martinez-Alier, 1998). Watts (2000) states that the aim of political ecology is to understand the complex relations between nature and society through a careful analysis of what one might call the forms of access and control over resources and their implications for environmental health and sustainable livelihoods (Watts, 2000). Stott and Sullivan (2000) argue that political ecology identified the political circumstances that forced people into activities which caused environmental degradation in the absence of alternative possibilities ... involved the query and reframing of accepted environmental narratives, particularly those directed via international environment and development discourses (Stott & Sullivan, 2000). Forsyth (2001) suggests that the aim of a realist political ecology is to understand the political ramifications of environmental degradation, but in a way that acknowledges the social and political construction of definitions of degradation (Forsyth, 2001). Robbins (2012) defines political ecology as 'empirical, research-based explorations to explain linkages in the condition and change of social/environmental systems, with explicit consideration of relations of power' (Robbins, 2012). Benjaminsen and Svarstad (2021) suggest that political ecology is about how the natural environment is managed by people. Local situations are studied in the light of national and global influences, and political ecology provides a critical alternative to other ways of studying environmental issues. This implies that political ecologists examine power relationships and question mainstream claims about environment and development that often are taken for granted (Benjaminsen & Svarstad, 2021).

The interactions between society and the environment occur at multiple levels. The complexities involved in the society-environment interactions are often influenced by multiple interests of the diverse set of state and non-state actors. However, it often results in unequal distribution of environmental resources that can benefit and/or deprive an individual or a group of stakeholders involved, depending on their ability to influence the decision-making processes. In its wider definition, the discourse of political ecology deals with the asymmetric power relations between the rich and the poor about their access to and control over natural resources from the perspective of resource appropriation for profit making by the rich to sustenance of basic livelihood needs of the marginalized sections. Thus, the political ecology paradigm is primarily concerned with how unequal power relations influence the contested claims over natural resources in Third World countries such as India. It provides a robust framework to study the capitalistic processes that lead to degradation of natural resources such land, water, forests, pastures and fisheries, etc. and their impact on livelihoods of the marginalized communities particularly small peasants, indigenous forest dwellers, pastoralists and small-scale fishers, etc. In the context of mineral and forest resources, the analytical framework of political ecology studies how access, ownership and control over various natural resources such as minerals and forests are exercised, who are the powerful state and non-state actors involved in the process of resource appropriation, what their implicit interests are, how do various dominant socio-economic, political and neo-liberal narratives influence the decision-making process and eventually who gains and who loses in the multi-layered interactions and negotiations involved in the governance of mineral and forest resources. The political ecology perspective also provides important insights into the root causes of the resource conflict over access and ownership rights over these resources and the broader consequences of such resource conflict on livelihoods of the marginalized tribal communities. Thus, the political ecology approach offers an alternative framing of the policy and governance problems from the perspectives of the marginalized sections of the society. It provides holistic understandings of the processes of resource appropriation, environmental degradation and the overall outcomes of such regressive transformation processes on livelihoods of the marginalized populations. Robbins (2012) argues that research tends to reveal winners and losers, hidden costs and the differential power that produces social and environmental outcomes. As a result, political ecological research proceeds from central questions, such as: what causes regional forest loss? Who benefits from wildlife conservation efforts and who loses? What political movements have grown from local land use transitions? In answering these, political ecologists follow a mode of explanation that evaluates the influence of variables acting at a number of scales, each nested within another, with local decisions influenced by regional polices, which are in turn directed by global politics and economics (Robbins, 2012). Perreault et al. (2015) suggest that political ecology is an explicitly normative intellectual project, which has from its beginning highlighted the struggles, interests and plight of marginalized populations: peasants, indigenous peoples, ethnic and religious minorities, women and the poor. In this sense, and in contrast to many other approaches of social and environmental analysis (e.g. cultural ecology, land use/land cover change analyses, etc.), political ecology is explicitly normative in its approach. Political ecologists thus seek not just to explain social and environmental processes, but to construct an alternative understanding of them, with an orientation towards social justice and radical politics (Perreault et al., 2015).

7.3 (Over)Exploitation of Natural Resources and Its Impact on Tribal Communities in Bastar

Bastar division of Chhattisgarh has some of India's most precious mineral resources in terms of iron ore deposits. Bailadila hills in Dantewada district are famous for its iron ores located in Kirandul and Bacheli complexes having a balance reserve of 378 million tonnes. National Mineral Development Corporation (NMDC Ltd.) has been engaged in open cast mining operations in the Bailadila hills since 1970s. A number of annual streams and rivers originate from these hills that merge into the Indravati River – the major river system which flows through Bastar before uniting with the Godavari river near Bhadrakali in Bhopalpatnam block of Bijapur district. The open cast mining operations in Bailadila hills have highly polluted most of these annual rivers and streams with heavy flow of iron metal effluents. For instance, the river that passes through Koter and Cherpal villages in Gangaloor area of Bijapur district often gets heavily polluted during monsoons. The tribal households of Cherpal and Koter most often lose their cattle who die due to consumption of contaminated water from the river. The water pollution caused due to open cast mining also leads to several adverse health impacts for the tribal populations such as skin diseases, etc. On the other hand, the state and a handful of mining companies including NMDC earn a huge profit margin from the mining operations through export of precious mineral resources from Bastar. However, very little is invested in the welfare of the tribal communities in the region. In recent years, crores of rupees have been pumped to the District Mineral Foundations (DMFs) from the profit of such mining operations in Bastar. These funds have been often invested in infrastructure development projects and meeting administrative expenses of senior government officials at the district level. Rather, this should have been ideally invested in programmes and projects related to livelihood, health and education of the tribal communities in the region for which it was originally meant. Thus, the tribal communities have largely remained deprived of the benefits of mining operations in Bastar and have been made to suffer from displacement and environmental pollution-related issues in the region.

Of late, NMDC has been involved in setting up a greenfield-integrated steel plant at Nagarnar located about 15 km from Jagdalpur town on the bank of Indravati River on Chhattisgarh-Odisha border. The Nagarnar steel plant has a capacity to produce up to 3 MTPA of processed iron with a total budget of Rs. 20,000 crore in 2009–2010. The steel plant which has been set up in the midst of a highly productive agriculture belt is at the centre of controversies in recent times leading to a number of human rights violation issues of the tribal communities, including displacement and land acquisition through illegal means. Also, the central government has decided to disinvest 51% of the share in favour of a private mining company in 2016. If the plant is completed sooner and a private mining company takes over of its operation, which is likely, the precious mineral resources of Bastar will be exploited at an alarming rate to generate profits for private companies. This will further deprive the local tribal populations of their rights of being benefitted from the mineral resources and force the tribal youth to join the Maoists force in absence of decent employment opportunities in the region.

The Rowghat mines located in Antagarh tehsil of Kanker district in North Bastar region have the second largest deposits of iron ore with an estimated reserve of 731.93 MT after the Bailadila hills in Chhattisgarh. Steel Authority of India (SAIL), which has made application to mine in F block of the Rowghat deposits that spreads across an area of 2,029 ha of forest land and possess an estimated 476.45 MT in order to supply the iron ore to its Bhilai Steel Plant has been granted conditional environmental clearance by the Ministry of Environment and Forests via the Supreme Court in 2008. Subsequently, the Government of Chhattisgarh has granted the mining licence to SAIL for mining in the F block deposit for a period of 20 years in 2009. The Rowghat mines were originally discovered by the Geological Survey of India in 1983 and since then ran into various issues including delay in environmental clearance by the Ministry of Environment and Forests, protests by local tribal communities and resistance from the Maoists. The Dalli-Rajahara–Jagdalpur rail line which is presently under construction will connect Rowghat to Dalli-Rajahara to be primarily used for transportation of the iron ore.[1]

The Rowghat hills are home to several floral and faunal species that harbour a unique ecosystem and highly rich in biodiversity in the region. The hills are also of important religious and cultural significance to the local tribal community of Maria Gond – a PVTG inhabiting the area. The Rowghat hills having rich forests with abundance of Sal, Teak and other native species of trees are the origin of many annual rivulets and small streams that provide vital ecosystem and economic services including water and NTFPs inevitable for sustenance of the lives and livelihoods of the local tribal communities. The forests are also part of an important wildlife corridor and habitat that support free movement of many wildlife species to the dense forests in Garhchiroli region of Maharashtra and help in natural genetic exchange of wild animals. It is highly likely that the mining activities in the Rowghat hills that have started since 2015 will cause irreversible damages to the rich biodiversity, lead to large scale felling of the trees, causing massive damages to the environment and destroy the socio-cultural and religious fabrics of the local tribal communities in the North Bastar region.

Bodhghat is a proposed hydroelectric project to be constructed on the Indravati River near Barsoor located about 100 km from the district headquarters at Jagdalpur in Bastar district. Originally cleared by the Planning Commission and inaugurated by the then late Prime Minister, Shri. Morarji Desai in 1979, the Bodhghat project proposes to construct a composite dam of 1,720 m length and 90 m height which will have four power generating units of 125 MW each. In fact, the proposed Bodhghat dam is the first one in a series of dams which have been planned to be constructed downstream of the Indravati River near Kurtu, Nugur, Bhopalpatnam and Ichhampali. The project in its original form will impact a total area of 13,783.14 ha of land that includes 5,704.33 ha of forest land displacing 10,000 tribals in 42 villages. According to a report submitted by the Wildlife Institute of India (WII) to the Ministry of Environment and Forests in 1990, it was suggested that the river Indravati and its tributaries have given rise to a unique Indravati reparian ecological systems comprises of Sal, Teak and Bamboo forests with patches of highly productive grassland which harbour several wild animals including the endangered wild buffaloes in the region. Moreover, a significant portion of the local populations which belong to Gond, Maria, Abujh Maria and Moria tribal communities, some of whom belong to PVTGs inhabit the area, have unique socio-cultural identities and livelihood patterns are highly dependent on the NTFPs collected from the forests for their livings.[2]

However, referring to the guidelines of the Forest Conservation Act, 1980, the Ministry of Environment and Forests constituted a working committee which visited the proposed dam site and held consultations way back in 1985. Based on the recommendations of the working committee and the basis of the report submitted by WII, the Ministry of Environment and Forests declined to give approval to transfer the 5,704.33 ha of forest land for non-forestry purpose required for construction of the proposed dam and cancelled the clearance issued in 1979. However, owing to the interferences of the then Union Minister of Environment and Forests who incidentally hailed from Chhattisgarh, the Ministry of Environment and Forests gave in-principle clearance for the forest land diversion in 2004. In an interesting development, the current state government in Chhattisgarh has shown its interest in resuming the construction of the proposed Bodhghat dam with an outlay of Rs. 20,000 crore and has presently commissioned Wapcos Ltd., a private infrastructure development company for initiating the survey work for the Bodhghat project. If undertaken, the project will have devastating effects on the fragile ecology of the region, severely destroy and fragment the natural habitats of wild buffaloes and other species of wildlife such as tigers, leopards and foxes in Bhairamgarh Wildlife Sanctuary and Indravati Tiger Reserve, two major protected areas located downstream of the Bijapur and Narayanpur districts in the Abhujamarh region by restricting their movement through blockage in the wildlife corridor that connects the forest patches of Garhchiroli in Maharashtra.

7.4 Wither NTFP-Based Livelihoods of Tribal Communities in Bastar

The tribal households are highly dependent on Non-Timber Forest Products (NTFPs) for their livelihoods in Bastar. The seasonality of NTFP collection approximately spans over four months from late February to mid of June in Bastar. Tendu leaves, Mahua flower, wild mango, Tamarind and Chironji are the major NTFPs collected by the tribal households in the region. Mahua flower is collected in February–March followed by wild mango in April–May. Tendu leaves are harvested during the first week of May. A household survey conducted by the author in select villages of Bijapur district in 2017 found that, on average, a household collects about 150 kg of Mahua flower over a period of about 25 days during the months of March and April. It was found that about 25 kg of Mahua flower is kept for self-consumption and the rest is sold in the local market (haat) to village-level traders (kuchia) for Rs. 10/- per kg. Thus, by selling Mahua flowers, a household earns up to Rs. 500/- in a year. Similarly, wild mango is collected during the months of May and June for about five days during the season. A household collects about 15 kg of wild mango which is sold in the local haats at Rs. 50/- per kg, thus earning up to Rs. 500/- during the season. Tamarind, Mahua seeds and Chironji are the other NTFPs collected in small quantities that are kept mostly for self-consumption.

Tendu leaf (*patta*) is the major NTFP harvested by the tribal households in Bastar. Tendu leaf is auctioned by the Forest Department through the Chhattisgarh State Minor Forest Produce Cooperative Federation (CGMFPFED) to private traders from Andhra Pradesh and Telangana. These private traders procure the Tendu leaf from the primary collectors during the first week of May each year. Both tribal men and women along with children are engaged in the collection of Tendu leaf. They often go to harvest Tendu leaf early in the morning and return with small headloads by noon. They pluck tender and good quality Tendu leaves mostly from regenerated trees in the nearby forests. The distance covered for the collection may range up to 3 to 5 km inside the forests depending on the availability of Tendu trees. The Tendu leaf collection season is over in a week's time. Upon their return post collection, women, men and children are engaged in bundling of the leaves. The leaves are then counted and kept in a bundle (*gaddi*) of 50 leaves each having 25 leaves stalked in opposite manner. A household harvests about 300 to 500 bundles of Tendu leaves in a day which depends on the number of persons involved and the hours of collection done in a day. For each 100 bundles of leaves (i.e. 50 × 100 = 5,000 Tendu leaves), a household is paid Rs. 220 to 230 including the bonus declared by the state government which is revised every year. On average, a tribal household harvests about a total 2,000 bundles of Tendu leaves for 7 days in the first week of May. Thus, they earn about Rs. 4,500/- through sale of Tendu leaves including the bonus received from the state government. The net amount is credited to their bank accounts through online transfer.

Apart from Tendu leaf, most of the other NTFPs such as Mahua flower are sold by the households in local haats. Households that are in dire need of cash sell these NTFPs to local traders at a very low price. Most of these NTFPs are sold within three months of harvesting. The prices of NTFPs are highly fluctuating with no specific trend over the past years. Selling of NTFPs tends to be extremely low remunerative activity. The household survey found that out of the total income earned from the sale of NTFPs, i.e., Rs. 5,200/- in a year by a household, a larger share comes from sale of Tendu leaf that contributes about Rs. 4,500/- (86%). However, income from Mahua flower (6%) and wild mango (7%) put together comprises less than 15% of the total income from NTFPs in a year. It was found that subsistence rainfed agriculture coupled with seasonal collection of NTFPs form the primary source of livelihoods of the tribal households in Bastar. However, degradation of the forest resources coupled with corruption in Tendu leaf auctioning process, unscientific harvesting practices and lack of storage and processing facilities have highly affected the NTFP-based livelihoods of the tribal households in Bastar. Also, lack of local employment opportunities during lean season forces the tribal households to go for distress sale of their produce to village-level traders in local *haat*. These petty traders and middlemen exploit the poor tribals by offering them very low price for their produce. Thus, the tribal households earn a meagre income from sale of the NTFPs in the region. However, in absence of a secure source of income, earnings from NTFPs significantly contribute to their livelihoods and plays a crucial role in the household economy in Bastar.

7.5 Development as the Panacea to Left-Wing Extremism in Bastar

More than 500 villages are currently part of the Red Corridor in Bastar region. Most of these villages are inhabited by marginalized tribal communities. Incidentally, many of these villages were highly affected by the Salwa Judum campaign during 2005–2007. However, most of the tribal households in these villages suffered considerable loss to their lives and livelihoods during the Salwa Judum movement. In fear of the violence perpetrated by the Salwa Judum activists and counter-violence of the Maoists, many of them have migrated to the neighbouring states of Andhra Pradesh and Telangana, some of them have fled to the dense forests and a few stayed in the temporary road-side relief camps or porta cabins, as they is locally known. Some of the affected households have lost their relatives, their food grains, cattle, poultry birds and mud huts were burnt by the Salwa Judum activists. Tribal women and girls have been sexually assaulted and brutally killed during the Salwa Judum movement, leaving them to be in inhuman conditions and suffering from severe physical miseries and mental trauma (Sundar, 2006). However, the fear and extreme violence that the Salwa Judum activists inflicted on them have left lasting impressions in their lives. As Guha (2007) argues that a democratic state should fight the rise of Maoist extremism by bringing the fruits of development to the *adivasi* and by prompt and effective police action. However, the policies currently being followed by the government of India are the antithesis of what one would prescribe. Instead of making tribals partners in economic development, they marginalize them further (Guha, 2007).

Of the 500 odd villages in the Red Corridor, the state has its presence in only about 50% of the villages through the district and block administrations in the entire Bastar region. As far as access to these villages by government officials is concerned, primary health workers and school teachers are the ones who regularly visit the interior villages. However, officials of other government line departments rarely visit the villages. Also concerned officials of forest and revenue departments are very rare to be seen let alone performing their duties in these villages. Central and state governments sponsored programmes and schemes such as Mahatma Gandhi National Rural Employment Guarantee Scheme (MGNREGS), Aspirational District Programme (ADP), National Rural Livelihood Mission (NRLM), Pradhan Mantri Aawas Yojana (PMAY) and Swatch Bharat Mission (SBM-G) are the major flagship programmes being implemented in the villages through Gram Panchayats. However, the target-oriented approach, lack of proper participatory planning, execution of low-quality work, high level of corruption by the Panchayat functionaries and concerned government officials and commission to the Maoists from the funds of these programmes fail to deliver desired benefits to the intended beneficiaries leaving the tribal households to be highly vulnerable and live in extremely miserable conditions in Bastar. Banerjee and Saha (2010) argue that underdevelopment or backwardness in the Indian context can largely be understood in terms of the development paradigm followed by successive governments. Socio-economic deprivation and exclusion have resulted in the growth of the Maoists in the backward areas of the country. The levels of rural distress can be tackled – the proper working of the NREGA seems to be a small but necessary step in that direction (Banerjee & Saha, 2010).

Similarly, Ghatak and Vanden Eynde (2017) suggest that development interventions are one of the best strategies to counter the Maoist insurgency in India. The Indian state urgently needs to adopt a multi-pronged strategy to ensure the well-being of the tribal households in Bastar. It needs to implement various development programmes and schemes to their fullest potential. Successful implementation of central government's flagship programmes such as MGNREGS, NRLM and FRA, 2006 will result significant positive outcomes. The district administration should strategically plan for massive public works under MGNREGS that needs to be carefully executed as there is higher demand from the existing workforce and from the migrant labourers who have returned to their villages during the COVID-19 pandemic. Participatory planning and implementation strategies for integrated natural resource management (INRM)-based activities should be adopted for the creation of durable assets such as percolation tanks, farm ponds, watershed development and protective irrigation facilities including dug wells in the villages. This will have long-term impacts on small and marginal tribal farmers by enhancing their income through increased agriculture productivity alongside providing wage employment opportunity to landless labourers and poorest tribal households in their villages discouraging them to depend on distress seasonal labour migration to earn their livings. The INRM works need to be supplemented by agriculture extension activities and horticulture-based interventions through convergence with concerned line departments of state government for maximum impact on the tribal farmers. NRLM is a potential scheme to mobilize tribal women into Self Help Groups (SHGs), imparting them basic financial literacy skills and providing them a platform to engage with formal financial institutions and various government schemes to initiate livelihood activities through micro-entrepreneurship development in the villages. Sustained training and capacity building of SHG members, linking them with banks for required credit, providing them funds through NRLM and convergence with other government programmes, handholding support on technical and marketing aspects of locally feasible on- and off-farm livelihood interventions in agriculture, animal husbandry, i.e., goat rearing and poultry, NTFPs and skill development-based livelihood promotion activities will ensure the socio-economic empowerment of tribal women in Bastar region. The district administration should also use the funds received through District Mineral Foundation (DMF) for improving the existing facilities of healthcare, education and livelihoods of local communities. Moreover, skill development and job placements of tribal youth are the other alternate livelihood options that the state government should implement on priority basis under Rural

Self-Employment Training Institute (RSETI) and Din Dayal Upadhyay Gramin Kaushalya Yojana (DDUGKY) in Bastar. More job-oriented vocational training programmes, practical orientation and industrial exposure, job placements for successful candidates, credit facilities through banks for initiating their own micro-enterprises or rural start-ups for generation of self-employment opportunities and career counselling services will motivate the tribal youth to join the mainstream and reduce their level of distress.

7.6 Strengthening Governance of Natural Resources in Bastar

The Forest Rights Act, 2006 (FRA) holds much promise in the region as there are many tribal households eligible for land titles under Individual Forest Right (IFR) that has been implemented to a certain extent in Bastar. However, the provision of Community Forest Rights (CFR) under FRA, 2006 has not been effectively implemented so far. Its implementation will provide justice to the marginalized tribal communities who have been struggling for legal recognition of their communal rights over forest resources – NTFPs in particular in Bastar. Effective implementation of various provisions of FRA would enable the marginalized tribal communities to assert their due rights over forest land and access to NTFPs, thereby ensuring their active participation in the resource governance process that has significant positive outcomes for their livelihoods in Bastar region. Their active participation in the forest governance process will also result in conservation of biodiversity which is on rapid decline through community-based conservation initiatives and use of their traditional knowledge in Bastar. Considering that Bastar comes under Schedule 5 Area as per the Indian Constitution, implementation of the Panchayats Extensions to the Scheduled Areas Act, 1996 (PESA) has much potential that has not been properly implemented so far in the entire Bastar region. Incidentally, it was the experience of the then collector of undivided Bastar, Dr. B.D. Sharma that led to the formulation of the PESA Act in 1996 advocating for greater autonomy of Schedule 5 Areas through active participation of tribal communities in the governance process. Its effective implementation will empower Gram Sabhas, thereby providing opportunities to the marginalized tribal communities to widely participate in the governance systems and protect their rights over access to and sustainable use of various forest resources. As the tribal households considerably depend on NTFPs for their livelihoods in Bastar, the state government should invest substantial resources in sustainable harvesting, storage, processing, value addition and marketing of various NTFPs, organize the primary collectors, most of whom are women into cooperatives or SHGs for promoting NTFP-based micro-enterprises at the village clusters. This will help the tribal women to earn a decent living through NTFP-based income-generation activities, ensure sustainable use of the NTFP resources through conservation and also enable the government to earn revenues from the NTFP trade in Bastar.

7.7 Conclusion

By using the analytical framework of political ecology, the paper attempted to provide a micro perspective of the contemporary issues of over-exploitation of natural resources and its impact on the tribal communities through ethnographic research in Bastar, Chhattisgarh. It finds that, despite abundant natural resources, the tribal populations live in extreme poverty and confront myriad socio-economic hardships in their day-to-day life in Bastar today. The impacts of mining and other development projects have led to dispossession of the tribal communities, leading to devastating effects on their livelihoods and on the environment in the entire region. The mining projects have largely benefitted the state and the mining corporations through over-exploitation of precious iron ore deposits in Bastar. However, very little has been invested in improving the socio-economic conditions of the tribal communities in the region. The ongoing resource conflict between the Left-Wing Extremists and the Indian state has seemingly paralyzed the day-to-day lives of the tribals, perpetuating extreme violence in Bastar, particularly over the past two decades. The ensuing bloodshed is unlikely to have any significant positive outcomes for the tribal households, a large majority of whom have been highly affected by it. Moreover, the recent COVID-19 pandemic has also added much desperation to the existing predicaments of the tribal communities in Bastar. It is suggested that effective implementation of development programmes such as MGNREGS, NRLM, FRA, 2006 and PESA Act, 1996 be applied to their fullest potential, and that the active participation of the tribal communities is encouraged in the governance process, protecting their rights over access to natural resources, substantial investments in healthcare, quality education and local infrastructure development facilities, creating decent employment opportunities for the tribal youth, improving the livelihoods of small and marginal tribal farmers and women through suitable livelihood interventions in agriculture, and allied sectors including NTFPs will provide long-term solutions to the existing woes of the marginalized tribal communities and ensure peace and well-being in Bastar.

Notes

1 https://ejatlas.org/conflict/rowghat-iron-ore-mines-chhattisgarh-india. Accessed on 20 June 2021.
2 www.sanctuaryasia.com/component/content/article/119-campaigns-archive /682-bodhghat-damning-the-indravati-.html. Accessed on 20 June 2021.

References

Banerjee, K., & Saha, P. (2010). The NREGA, the maoists and the developmental woes of the Indian State. *Economic and Political Weekly, XLV*(28), 42–47.

Benjaminsen, T. A., & Svarstad, H. (2021). *Political ecology: A critical engagement with global environmental issue* (1st ed.). Palgrave Macmillan.

Blaikie, P., & Brookfield, H. (1987). *Land degradation and society*. London: Methuen.

Bryant, R. L. (1998). Power, knowledge and political ecology in the third world: A review. *Progress in Physical Geography, 22*(1), 79–94.

Bryant, R., & Bailey, S. (1997). *Third world political ecology*. London: Routledge.

Forsyth, T. (2001). Critical realism and political ecology. In A. Stainer & G. Lopez (Eds.), *After postmodernism: Critical realism?* (pp. 146–154). London: Athlone Press.

Ghatak, M., & Vanden Eynde, O. (2017). Economic determinants of the maoist conflict in India. *Economic and Political Weekly, LII*(39), 69–76.

Greenberg, J. B., & Park, T. K. (1994). Political ecology. *Journal of Political Ecology, 1*(1), 1–12. doi: https://doi.org/10.2458/v1i1.21154.

Guha, R. (2007). Adivasis, Naxalites and Indian democracy. *Economic and Political Weekly, 42*(32), 3305–3312.

Guha, R., & Martinez-Alier, J. (1997). *Varieties of environmentalism: Essays north and south*. London: Earthscan.

Perreault, T., Bridge, G., & McCarthy, J. (2015). *The Routledge handbook of political ecology*. London and New York: Routledge.

Robbins, P. (2012). *Political ecology: A critical introduction* (2nd ed.). Chichester: Wiley & Sons.

Stott, P., & Sullivan, S. (2000). *Political ecology: Science, myth and power*. London: Arnold.

Sundar, N. (2006). Bastar, maoism and Salwa Judum. *Economic and Political Weekly, 41*(29), 3187–3192.

Watts, M. J. (2000). Political ecology. In E. Sheppard & T. Barnes (Eds.), *A companion to economic geography* (pp. 257–274). Oxford: Blackwell.

Index